Simulating Data with SAS®

Rick Wicklin

support.sas.com/bookstore

The correct bibliographic citation for this manual is as follows: Wicklin, Rick. 2013. *Simulating Data with SAS®*. Cary, NC: SAS Institute Inc.

Simulating Data with SAS®

Contents

Acknowledgments

I would like to thank Robert Rodriguez and Phil Gibbs for pointing out the need for a book about simulating data in SAS. "Simulation" is a vast topic, and early discussions with them helped me to whittle down the possible topics. Bob and Maura Stokes provided many opportunities for me to develop this material by inviting me to present papers and workshops at conferences. My supervisors at SAS fully supported me as I prepared for and participated in these conferences.

I thank the many SAS users who encouraged me to write a book that emphasizes the practical side of simulation. Discussions with SAS users helped me to determine what topics are of practical importance to statisticians and analysts in business and industry.

I thank my colleagues at SAS from whom I have learned many statistical and programming techniques. Special thanks to Randy Tobias, who always provides sound advice and statistical wisdom for my naive questions. Thanks also to Tim Arnold and Warren Kuhfeld for their 'saslatex' documentation system that automatically produced all tables and graphs in this book from the programs that appear in the text.

I thank my editor, John West, and the other employees at SAS Press for their work producing and promoting the book. I thank two reviewers, Clement Stone and Bob Pearson, who provided insightful comments about the book's content and organization.

Thanks to several colleagues and friends who read and commented on early drafts of this book. This includes the following individuals: Rob Agnelli, Jason Brinkley, Tonya Chapman, Steve Denham, Bruce Elsheimer, Betsy Enstrom, Phil Gibbs, Emily Lada, Pushpal Mukhopadhyay, Bill Raynor, Robert Rodriguez, Jim Seabolt, Udo Sglavo, Ying So, Jill Tao, Randy Tobias, Ian Wakeling, Donna Watts, and Min Zhu.

Finally, I would like to thank my wife, Nancy, for her constant support, and my parents for instilling in me a love of learning.

Part I

Essentials of Simulating Data

Chapter 1
Introduction to Simulation

Contents

1.1 Overview of Simulation of Data

There are many kinds of simulation. Climate scientists use simulation to model the interactions between the earth's atmosphere, oceans, and land. Astrophysicists use simulation to model the evolution of galaxies. Biologists use simulation to model the spread of epidemics and the effects of vaccination programs. Engineers use simulation to study the safety and fuel efficiency of automobile and airplane designs. In these simulations of physical systems, scientists model reality and use a computer to study the model under various conditions.

Statisticians also build models. For example, a simple model of human height might assume that height is normally distributed in the population. This is a useful model, but it turns out that human heights are not actually normally distributed (Schilling, Watkins, and Watkins 2002). Even if you restrict the data to a single gender, there are more very tall and very short people than would be expected from a normal distribution of heights.

If a set of data is only approximately normal, what does that mean for statistical tests that assume normality? If you compute a t test to compare the means of two groups—a test that assumes that the two underlying populations are normally distributed—how sensitive is your conclusion to the actual population distribution? If the populations are slightly nonnormal, does that invalidate the t test? Or are the results fairly robust to deviations from normality?

One way to answer these questions is to simulate data from nonnormal populations. If you construct a distributional model, then you can generate random samples from the model and examine how the t test performs on the simulated data. Simulation gives you complete control over the characteristics of the population model from which the (simulated) data are drawn.

Simulating data is also useful for comparing two different statistical techniques. Perhaps Technique A performs better on skewed data than Technique B. Perhaps Technique B is more robust to the presence of outliers. To a practicing statistician, this kind of information is quite valuable. As Gentle (2009, p. xi) says, "Learning to simulate data with given characteristics means that one understands those characteristics. Applying statistical methods to simulated data ... helps us better to understand those methods and the principles underlying them."

This book is about simulating data in SAS software. This book demonstrates how to generate observations from populations that have specified statistical characteristics. In this book, the phrases "simulating data," "generating a random sample," and "sampling from a distribution" are used interchangeably.

A large portion of this book is about learning how to construct statistical models (distributions) that have certain statistical properties. Skewed distributions, fat-tailed distributions, bimodal distributions—these are a few examples of models that you can construct by using the techniques in this book. This book also presents techniques for generating data from correlated multivariate distributions. Each technique is accompanied by a SAS program.

Although this book uses statistics, its audience is not limited to statisticians. It is a how-to book for statistical programmers who use SAS software and who want to simulate data efficiently.

In short, this book describes how to write SAS programs that simulate data with a wide range of characteristics, and describes how to use that data to understand the performance and applicability of statistical techniques.

1.2   The Goal of This Book

The goal of this book is to provide tips, techniques, and examples for efficiently simulating data in SAS software.

Data simulation is a fundamental technique in statistical programming and research. To evaluate statistical methods, you often need to create data with known properties, both random and nonrandom. This book contains more than one hundred annotated programs that simulate data with specified characteristics. You can use simulated data to estimate the probability of an event, to estimate the sampling distribution of a statistic, to estimate the coverage probabilities of confidence intervals, and to evaluate the robustness of a statistical test.

Some programs are presented in two forms, first by using the DATA step and then again by using the SAS/IML language. By presenting the same algorithm in two different ways, the novice SAS/IML programmer can learn how to write simulations in the SAS/IML language. Later chapters that discuss multivariate simulation use the SAS/IML language heavily. If you are serious about simulation, you should invest the time to learn how to use the SAS/IML language efficiently.

Although this book covers many standard examples of data distributions, there are many other examples that are not covered. However, many techniques that are described in this book are generally applicable. For example, Section 7.5 describes how to generate random samples from a mixture of normal distributions. The same technique can be used to simulate data from a mixture of other distributions.

The book also includes more than 100 exercises. Many exercises extend the results of a section to other distributions or to related problems. The exercises provide practical programming problems that encourage you to master the material before moving on to the next section. Most exercises will take five to 15 minutes of programming. As Gentle (2009, p. xiii) says, "Programming is the best way to learn programming." Solutions to selected exercises are available from the book's Web site. See `support.sas.com/wicklin`.

1.3 Who Should Read This Book?

The audience for this book is statisticians, analysts, programmers, graduate students, and researchers who use SAS to analyze and model data.

This book assumes that you are familiar with SAS programming concepts such as missing values, formats, and the SAS DATA step. This book uses simple macro statements such as the %LET statement, but does not include sophisticated macro techniques.

This book presumes familiarity with basic statistical ideas that you might encounter when you use the FREQ, UNIVARIATE, CORR, TTEST, and GLM procedures. This book discusses random variables, quantiles, distributions, and regression. The chapters of the book that deal with multivariate distributions presume familiarity with concepts of computational linear algebra, such as matrix decompositions. There are also sections of the book that discuss various topics in regression analysis and distributional modeling.

1.4 The SAS/IML Language

The DATA step is sufficient for simulating data from simple univariate distributions. However, for simulating from more complicated distributions, SAS/IML software is essential.

IML stands for "interactive matrix language." The SAS/IML language enables you to implement custom algorithms that use vectors and matrices. The SAS/IML language contains hundreds of built-in functions and subroutines, and you can also call hundreds of functions in Base SAS software.

To learn how to write efficient programs in the SAS/IML language, see Wicklin (2010). The Web site for that book (`support.sas.com/wicklin`) has a "Getting Started" chapter that is available as a free Web download. Furthermore, the SAS/IML language is used and discussed frequently on the author's blog, *The DO Loop*, which can be read at `blogs.sas.com/content/iml`. Statistical simulation is a frequent topic on the blog.

This book explains SAS/IML statements that are not obvious. Readers who have previous matrix programming experience with MATLAB or the R language should be able to read the SAS/IML programs in this book. The three languages are similar in syntax.

You need a license for the SAS/IML product in order to run the IML procedure. To see whether PROC IML is licensed at your site, submit the following program:

```
proc product_status;
run;
```

If SAS/IML is listed in the SAS log, then SAS/IML software is licensed for your site.

To see if PROC IML is installed at your site, submit the following program:

```
proc iml;
quit;
```

If the SAS log contains the message ERROR: Procedure IML not found, then SAS/IML software is not installed. If the SAS log contains the message IML Ready, then SAS/IML is installed.

If SAS/IML software is licensed but not installed, ask your SAS administrator to install it.

1.5   Comparing the DATA Step and SAS/IML Language

Many data analysts use the DATA step as their primary programming tool. The DATA step is adequate for simulating univariate data from many distributions and for simulating uncorrelated multivariate data. However, simulating multivariate correlated data (and even data from complicated univariate distributions) is much easier if you use matrix-vector computations.

Conceptually, there are two main differences between the DATA step and a SAS/IML program. First, a DATA step implicitly loops over all observations; a typical SAS/IML program does not. Second, the fundamental unit in the DATA step is an observation; the fundamental unit in the SAS/IML language is a matrix.

The syntax of the SAS/IML language has much in common with the DATA step: neither language is case sensitive, variable names can contain up to 32 characters, and statements must end with a semicolon. Furthermore, the syntax for control statements such as the IF-THEN/ELSE statement and the iterative DO statement is similar for both languages. The two languages use the same symbols to test a quantity for equality (=) and inequality (^=), and to compare quantities (for example, <=). The SAS/IML language enables you to call the same mathematical functions provided in the DATA step, such as LOG, EXP, SQRT, CEIL, and FLOOR, except that the SAS/IML versions act on vectors and matrices.

SAS/IML software is intended for statistical computing, and writing a simulation in the SAS/IML language is usually more compact than writing the equivalent simulation in the DATA step. Furthermore, because the SAS/IML language keeps data in RAM (whereas the DATA step writes data sets), the SAS/IML language offers excellent performance for simulations in which the simulated data fit in memory and for which the computations can be *vectorized*.

A computation is vectorized if it consists of a few executable statements, each of which operates on a fairly large quantity of data, usually a matrix or a vector. A program in a matrix-vector language is more efficient when it is vectorized because most of the computations are done in a low-level language such as C. In contrast, a program that is not vectorized requires many calls that transfer small amounts of data between the high-level program interpreter and the low-level computational code. To vectorize a program in a matrix-vector language, take advantage of built-in functions and linear algebra operations. Avoid loops that access individual elements of matrices.

1.6   Overview of This Book

This book consists of four parts. The first part introduces essential concepts. It shows you how to use SAS software to simulate data from frequently used distributions, and how to compute useful related quantities. In addition to the current chapter, this part of the book contains the following chapters:

Chapter 2: Simulate univariate samples from common discrete and continuous distributions.

Chapter 3: Compute basic quantities in SAS software that are essential for simulating data.

The second part of the book describes how to use simulated data to examine the sampling distribution of statistics and to evaluate statistical techniques. This part of the book contains the following chapters:

Chapter 4: Use simulated data to estimate the sampling distributions of basic statistics such as the mean, median, and Pearson correlations.

Chapter 5: Use simulation to evaluate statistical techniques. Examples include using simulation to investigate the coverage probability of a confidence interval, to estimate p-values, and to estimate the power of a t test.

Chapter 6: Develop strategies for efficient and effective simulation in SAS software.

The third part of the book describes advanced simulation of univariate and multivariate data. It also describes how to construct covariance matrices that are often needed for simulation studies. This part of the book contains the following chapters:

Chapter 7: Develop advanced techniques in univariate simulation, including sampling from mixture distributions, acceptance-rejection sampling, and inverse CDF sampling.

Chapter 8: Simulate data from basic multivariate distributions.

Chapter 9: Simulate multivariate data with special structure, such as multivariate binary variables, ordinal variables, and data from copulas.

Chapter 10: Simulate correlation and covariance matrices with known properties and structure, such as Toeplitz or AR(1) structure. The chapter shows how to find the nearest correlation matrix to an estimate that is not positive semidefinite.

The fourth part of the book shows how to use simulation in statistical modeling. This part of the book contains the following chapters:

Chapter 11: Simulate data from a variety of basic regression models, such as linear models with continuous and classification variables.

Chapter 12: Simulate data from generalized linear models, mixed models, and models in survival analysis.

Chapter 13: Simulate data from time series.

Chapter 14: Simulate data from spatial models.

Chapter 15: Use bootstrap methods to resample from the data that you want to simulate.

Chapter 16: Use moment matching and moment-ratio diagrams to simulate data that have properties similar to a given set of real data.

This book also includes an appendix that provides additional background and details about programming in the SAS/IML language.

1.7   Obtaining the Programs Used in This Book

This book was developed using SAS 9.3M2, which is the second maintenance release of SAS 9.3 and was released in August 2012. This release includes SAS/IML 12.1 software. When a SAS/IML 12.1 feature is used, the book also describes how to obtain the same result by using SAS/IML 9.3 software.

The SAS programs in this book are available as a free download from the book's Web site: `support.sas.com/wicklin`.

Of particular interest are dozens of simulation algorithms that the author implemented in the SAS/IML language. You can use these functions in your own simulation studies by doing the following:

- Download the zip file that includes the programs for this book.

- The zip file includes the file *SimulatingData.sas*, which contains the SAS/IML functions.

- Save *SimulatingData.sas* to a convenient location, such as
  `C:\Users\<userid>\Documents\My SAS Files`.

Whenever you want to use the SAS/IML functions, submit the following statement prior to calling PROC IML:

```
%include "C:\Users\<userid>\Documents\My SAS Files\SimulatingData.sas";
```

The statement defines all of the SAS/IML functions and stores them in a library. To use a function, you can load it by name from within PROC IML. To load all of the functions, run the following SAS/IML statement:

```
load module=_all_;
```

1.8   Specialized Simulation Tools in SAS Software

This book uses the SAS DATA step and SAS/IML software to simulate data. The book also uses the SIMNORMAL and SIM2D procedures in SAS/STAT software, and the COPULA procedure in SAS/ETS software.

SAS software contains other specialized simulation tools that are not covered in this book, including the following:

- SAS Simulation Studio, which is part of the SAS/OR product, enables you to build discrete event simulations such as those that arise in queuing theory.

- The MCMC procedure in SAS/STAT software is a general-purpose Markov chain Monte Carlo procedure. This procedure enables you to use simulation to fit Bayesian models. You can also use the MCMC procedure to perform direct sampling.

- PROC MODEL in SAS/ETS software enables you to perform Monte Carlo simulation of time series models.

If you are interested in these topics, it is worthwhile to learn how to use these specialized tools. Although some of the techniques in this book can be applied to topics such as Monte Carlo integration and Bayesian analysis, using a specialized tool such as PROC MCMC is easier and more efficient than using general-purpose techniques.

1.9   References

Gentle, J. E. (2009), *Computational Statistics*, New York: Springer-Verlag.

Schilling, M. F., Watkins, A. E., and Watkins, W. (2002), "Is Human Height Bimodal?" *American Statistician*, 56, 223–229.
  URL http://www.jstor.org/stable/3087302

Wicklin, R. (2010), *Statistical Programming with SAS/IML Software*, Cary, NC: SAS Institute Inc.

Chapter 2
Simulating Data from Common Univariate Distributions

Contents

2.1 Introduction to Simulating Univariate Data

There are three primary ways to simulate data in SAS software:

- Use the DATA step to simulate data from univariate and uncorrelated multivariate distributions. You can use the RAND function to generate random values from more than 20 standard univariate distributions. You can combine these elementary distributions to build more complicated distributions.

- Use the SAS/IML language to simulate data from many distributions, including correlated multivariate distributions. You can use the RANDGEN subroutine to generate random values from standard univariate distributions, or you can use several predefined modules to generate data from multivariate distributions. You can extend the SAS/IML language by defining new functions that sample from distributions that are not built into SAS.

- Use specialized procedures in SAS/STAT software and SAS/ETS software to simulate data with special properties. Procedures that generate random samples include the SIMNORMAL, SIM2D, and COPULA procedures.

This chapter describes the two most important techniques that are used to simulate data in SAS software: the DATA step and the SAS/IML language. Although the DATA step is a useful tool for simulating univariate data, SAS/IML software is more powerful for simulating multivariate data. To learn how to use the SAS/IML language effectively, see Wicklin (2010).

Most of the terminology in this book is standard. However, a term that you might not be familiar with is the term *random variate*. A random variate is a particular outcome of a random variable (Devroye 1986). For example, let X be a Bernoulli random variable that takes on the value 1 with probability p and the value 0 with probability $1 - p$. If you draw five observations from the probability distribution, you might obtain the values $0, 1, 1, 0, 1$. Those five numbers are random variates. This book also uses the terms "simulated values" and "simulated data." Some authors refer to simulated data as "fake data."

2.2   Getting Started: Simulate Data from the Standard Normal Distribution

To "simulate data" means to generate a random sample from a distribution with known properties. Because an example is often an effective way to convey main ideas, the following DATA step generates a random sample of 100 observations from the standard normal distribution. Figure 2.1 shows the first five observations.

```
data Normal(keep=x);
call streaminit(4321);              /* Step 1 */
do i = 1 to 100;                    /* Step 2 */
   x = rand("Normal");             /* Step 3 */
   output;
end;
run;

proc print data=Normal(obs=5);
run;
```

Figure 2.1 A Few Observations from a Normal Distribution

Obs	x
1	1.24067
2	-0.53532
3	-1.01394
4	0.68965
5	-0.32458

The DATA step consists of three steps:

1. Set the seed value with the STREAMINIT function. Seeds for random number generation are discussed further in Section 3.3.

2. Use a DO loop to iterate 100 times.

3. For each iteration, call the RAND function to generate a random value from the standard normal distribution.

If you change the seed value, you will get a different random sample. If you change the number 100, you will get a sample with a different number of observations. To get a nonnormal distribution, change the name of the distribution from "Normal" to one of the families listed in Section 2.7. Some distributions, including the normal distribution, include parameters that you can specify after the name.

2.3 Template for Simulating Univariate Data in the DATA Step

It is easy to generalize the example in the previous section. The following SAS pseudocode shows a basic template that you can use to generate N observations with a specified distribution:

```
%let N = 100;                  /* size of sample                    */

data Sample(keep=x);
call streaminit(4321);         /* or use a different seed           */
do i = 1 to &N;                /* &N is the value of the N macro var */
   /* specify distribution and parameters */
   x = rand("DistribName", param1, param2, ...);
   output;
end;
run;
```

The simulated data are written to the Sample data set. The macro variable **N** is defined in order to emphasize the role of that parameter. The expression **&N** is replaced by the value of the macro parameter (here, 100) before the DATA step is run.

The (pseudo) DATA step demonstrates the following steps for simulating data:

1. A call to the STREAMINIT subroutine, which specifies the seed that initializes the random number stream. When the argument is a positive integer, as in this example, the random sequence is reproducible. If you specify 0 as the argument, the random number sequence is initialized from your computer's internal system clock. This implies that the random sequence will be different each time that you run the program. Seeds for random number generation are discussed in Section 3.3.

2. A DO loop that iterates N times.

3. A call to the RAND function, which generates one random value each time that the function is called. The first argument is the name of a distribution. The supported distributions are enumerated in Section 2.7. Subsequent arguments are parameter values for the distribution.

2.4 Simulating Data from Discrete Distributions

When the set of possible outcomes is finite or countably infinite (like the integers), assigning a probability to each outcome creates a *discrete probability distribution*. Of course, the sum of the probabilities over all outcomes is unity.

The following sections generate a sample of size $N = 100$ from some well-known discrete distributions. The code is followed by a frequency plot of the sample, which is overlaid with the exact probabilities of obtaining each value. You can use PROC FREQ to compute the empirical distribution of the data; the exact probabilities are obtained from the probability mass function (PMF) of the distribution. Section 3.4.2 describes how to overlay a bar chart with a scatter plot that shows the theoretical probabilities.

2.4.1 The Bernoulli Distribution

The Bernoulli distribution is a discrete probability distribution on the values 0 and 1. The probability that a Bernoulli random variable will be 1 is given by a parameter, $p, 0 \le p \le 1$. Often a 1 is labeled a "success," whereas a 0, which occurs with probability $1 - p$, is labeled a "failure."

The following DATA step generates a random sample from the Bernoulli distribution with $p = 1/2$. If you identify $x = 1$ with "heads" and $x = 0$ with "tails," then this DATA step simulates $N = 100$ tosses of a fair coin.

```
%let N = 100;
data Bernoulli(keep=x);
call streaminit(4321);
p = 1/2;
do i = 1 to &N;
   x = rand("Bernoulli", p);              /* coin toss */
   output;
end;
run;
```

You can use the FREQ procedure to count the outcomes in this simulated data. For this sample, the value 0 appeared 52 times, and the value 1 appeared 48 times. These frequencies are shown by the bar chart in Figure 2.2. The expected percentages for each result are shown by the round markers.

Figure 2.2 Sample from Bernoulli Distribution ($p = 1/2$) Overlaid with PMF

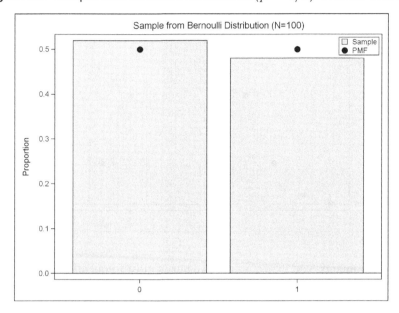

If X is a random variable from the Bernoulli distribution, then the expected value of X is p and the variance is $p(1 - p)$. In practice, this means that if you generate a large random sample from the Bernoulli distribution, you can expect the sample to have a sample mean that is close to p and a sample variance that is close to $p(1 - p)$.

2.4.2 The Binomial Distribution

Imagine repeating a Bernoulli trial n times, where each trial has a probability of success equal to p. If p is large (near 1), you expect most of the Bernoulli trials to be successes and only a few of the trials to be failures. On the other hand, if p is near $1/2$, you expect to get about $n/2$ successes.

The binomial distribution models the number of successes in a sequence of n independent Bernoulli trials. The following DATA step generates a random sample from the binomial distribution with $p = 1/2$ and $n = 10$. This DATA step simulates a series of coin tosses. For each trial, the coin is tossed 10 times and the number of heads is recorded. This experiment is repeated $N = 100$ times. Figure 2.3 shows a frequency plot of the results.

```
data Binomial(keep=x);
call streaminit(4321);
p = 1/2;
do i = 1 to &N;
   x = rand("Binomial", p, 10);      /* number of heads in 10 tosses */
   output;
end;
run;
```

Figure 2.3 Sample from Binomial Distribution ($p = 1/2, n = 10$) Overlaid with PMF

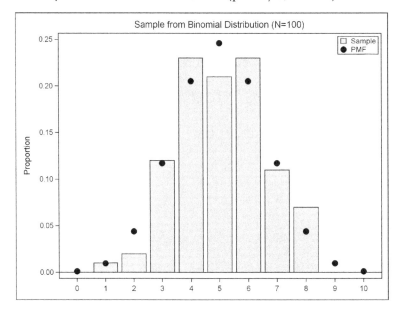

For this series of experiments, you expect to get five heads most frequently, followed closely by four and six heads. The expected percentages are indicated by the round markers. For this particular simulation, Figure 2.3 shows that four heads and six heads appeared more often than five heads appeared. The sample values are shown by the bars; the expected percentages are shown by round markers.

If X is a random variable from the binomial(p, n) distribution, then the expected value of X is np and the variance is $np(1 - p)$. In practice, this means that if you generate a large random sample from the binomial(p, n) distribution, then you can expect the sample to have a sample mean that is close to np.

Some readers might be concerned that the distribution of the sample shown in Figure 2.3 differs so much from the theoretical distribution of the binomial distribution. This deviation is not an indication that something is wrong. Rather, it demonstrates *sampling variation*. When you simulate data from a population model, the data will almost always look slightly different from the distribution of the population. Some values will occur more often than expected; some will occur less often than expected. This is especially apparent in small samples and for distributions with large variance. It is this sampling variation that makes simulation so valuable.

2.4.3   The Geometric Distribution

How many times do you need to toss a fair coin before you see heads? Half the time you will see heads on the first toss, one quarter of the time it requires two tosses, and so on. This is an example of a geometric distribution.

In general, the geometric distribution models the number of Bernoulli trials (with success probability p) that are required to obtain one success. An alternative definition, which is used by the MCMC procedure in SAS, is to define the geometric distribution to be the number of *failures* before the first success.

You can simulate a series of coin tosses in which the coin is tossed until a heads appears and the number of tosses is recorded. If p is the probability of tossing heads, then the following statement generates an observation from the Geometric(p) distribution:

```
x = rand("Geometric", p);        /* number of trials until success */
```

Figure 3.6 shows a graph of simulated geometric data and an overlaid PMF.

If X is a random variable from the geometric(p) distribution, then the expected value of X is $1/p$ and the variance is $(1 - p)/p^2$.

Exercise 2.1: Write a DATA step that simulates observations from a Geometric(0.5) distribution.

2.4.4   The Discrete Uniform Distribution

A Bernoulli distribution models two outcomes. You can model situations in which there are multiple outcomes by using either the discrete uniform distribution or the "Table" distribution (see the next section).

When you toss a standard six-sided die, there is an equal probability of seeing any of the six faces. You can use the discrete uniform distribution to produce k integers in the range $[1, k]$. SAS does not have a built-in discrete uniform distribution. Instead, you can use the continuous uniform distribution to produce a random number u in the interval $(0, 1)$, and you can use the CEIL function to produce the smallest integer that is greater than or equal to ku.

The following DATA step generates a random sample from the discrete uniform distribution with $k = 6$. This DATA step simulates $N = 100$ rolls of a fair six-sided die.

```
data Uniform(keep=x);
call streaminit(4321);
k = 6;                                  /* a six-sided die       */
do i = 1 to &N;
   x = ceil(k * rand("Uniform"));       /* roll 1 die with k sides */
   output;
end;
run;
```

You can also simulate data with uniform probability by using the "Table" distribution, which is described in the next section.

To check the empirical distribution of the simulated data, you can use PROC FREQ to show the distribution of the x variable. The results are shown in Figure 2.4. As expected, each number 1, 2,..., 6 is generated about 16% of the time.

```
proc freq data=Uniform;
   tables x / nocum;
run;
```

Figure 2.4 Sample from Uniform Distribution ($k = 6$)

The FREQ Procedure

x	Frequency	Percent
1	15	15.00
2	18	18.00
3	15	15.00
4	12	12.00
5	22	22.00
6	18	18.00

2.4.5   Tabulated Distributions

In some situations there are multiple outcomes, but the probabilities of the outcomes are not equal. For example, suppose that there are 10 socks in a drawer: five are black, two are brown, and three are white. If you close your eyes and draw a sock at random, the probability of that sock being black is 0.5, the probability of that sock being brown is 0.2, and the probability of that sock being white is 0.3. After you record the color of the sock, you can replace the sock, mix up the drawer, close your eyes, and draw again.

The RAND function supports a "Table" distribution that enables you to specify a table of probabilities for each of k outcomes. You can use the "Table" distribution to sample with replacement from a finite set of outcomes where you specify the probability for each outcome. In SAS/IML software, you can use the RANDGEN or SAMPLE routines.

The following DATA step generates a random sample of size $N - 100$ from the "Table" distribution with probabilities $p = \{0.5, 0.3, 0.2\}$. You can use PROC FREQ to display the observed frequencies, which are shown in Figure 2.5.

```
data Table(keep=x);
call streaminit(4321);
p1 = 0.5; p2 = 0.2; p3 = 0.3;
do i = 1 to &N;
   x = rand("Table", p1, p2, p3);          /* sample with replacement */
   output;
end;
run;

proc freq data=Table;
   tables x / nocum;
run;
```

Figure 2.5 Sample from "Table" Distribution ($p = \{0.5, 0.2, 0.3\}$)

The FREQ Procedure

x	Frequency	Percent
1	48	48.00
2	21	21.00
3	31	31.00

For the simulated sock experiment with the given probabilities, a black sock (category 1) was drawn 48 times, a brown sock (category 2) was drawn 21 times, and a white sock was drawn 31 times.

If you have many potential outcomes, it would be tedious to specify the probabilities of each outcome by using a comma-separated list. Instead, it is more convenient to specify an array in the DATA step to hold the probabilities, and to use the OF operator to list the values of the array as shown in the following example:

```
data Table(keep=x);
call streaminit(4321);
array p[3] _temporary_ (0.5 0.2 0.3);
do i = 1 to &N;
   x = rand("Table", of p[*]);              /* sample with replacement */
   output;
end;
run;
```

The _TEMPORARY_ keyword makes `p` a temporary array that holds the parameter values. The elements of a temporary array do not have names and are not written to the output data set, which means that you do not need to use a DROP or KEEP option to omit them from the data set.

The "Table" distribution is related to the multinomial distribution, which is discussed in Section 8.2. If you generate N observations from the "Table" distribution and tabulate the frequencies for each category, then the frequency vector is a single observation from the multinomial distribution. Consequently, the "Table" and multinomial distributions are related in the same way that the Bernoulli and binomial distributions are related.

Exercise 2.2: Use the "Table" distribution to simulate rolls from a six-sided die.

2.4.6   The Poisson Distribution

Suppose that during the work day a worker receives email at an average rate of four messages per hour. What is the probability that she might get seven messages in an hour? Or that she might get only one message? The Poisson distribution models the counts of an event during a given time period, assuming that the event happens at a constant average rate.

For an average rate of λ events per time period, the expected value of a random variable from the Poisson distribution is λ, and the variance is also λ.

The following DATA step generates a random sample from the Poisson distribution with $\lambda = 4$. This DATA step simulates the number of emails that a worker receives each hour, under the assumption that the number of emails arrive at a constant average rate of four emails per hour. This experiment simulates $N = 100$ hours at work. The results are shown in Figure 2.6.

```
data Poisson(keep=x);
call streaminit(4321);
lambda = 4;
do i = 1 to &N;
   x = rand("Poisson", lambda);          /* num events per unit time */
   output;
end;
run;
```

Figure 2.6 Sample from Poisson Distribution ($\lambda = 4$) Overlaid with PMF

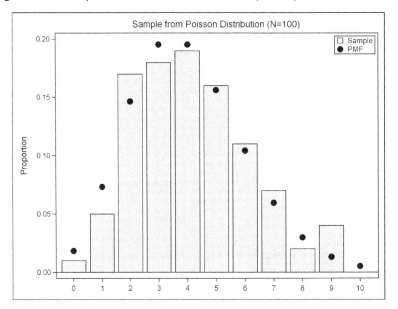

For the Poisson model, the worker can expect to receive four emails during a one-hour period about 20% of the time. The same is true for receiving three emails in an hour. She can expect to receive six emails during an hour slightly more than 10% of the time. These expected percentages are shown by the round markers. The "actual" number of emails received during each hour is shown by the bar chart for the 100 simulated hours. There were 18 one-hour periods during which the worker received three emails. There were 11 one-hour periods during which the worker received six emails.

Exercise 2.3: A negative binomial variable is defined as the number of failures before k successes in a series of independent Bernoulli trials with probability of success p. Define a trial as rolling a six-sided die until a specified face appears $k = 3$ times. Simulate 1,000 trials and plot the distribution of the number of failures.

2.5   Simulating Data from Continuous Distributions

When the set of possible outcomes is uncountably infinite (like an interval or the set of real numbers), assigning a probability to each outcome creates a *continuous probability distribution*. Of course, the integral of the probabilities over the set is unity.

The following sections generate a sample of size $N = 100$ from some well-known continuous distributions. Most sections also show a histogram of the sample that is overlaid with the probability density curve for the population. The probability density function (PDF) is described in Section 3.2. Section 3.4.3 describes how to create the graphs.

See Table 2.3 for a list of common distributions that SAS supports.

2.5.1   The Normal Distribution

The normal distribution with mean μ and standard deviation σ is denoted by $N(\mu, \sigma)$. Its density is given by the following:

$$f(x; \mu, \sigma) = \frac{1}{\sigma\sqrt{2\pi}} \exp\left(-\frac{(x-\mu)^2}{2\sigma^2}\right)$$

The *standard* normal distribution sets $\mu = 0$ and $\sigma = 1$.

Many physical quantities are modeled by the normal distribution. Perhaps more importantly, the sampling distribution of many statistics are approximately normally distributed.

Section 2.2 generated 100 observations from the standard normal distribution. Figure 2.7 shows a histogram of the simulated data along with the graph of the probability density function. For this sample, the histogram bars are below the PDF curve for some intervals and are greater than the curve for other intervals. A second sample of 100 observations is likely to produce a different histogram.

Figure 2.7 Sample from a Normal Distribution ($\mu = 0, \sigma = 1$) Overlaid with PDF

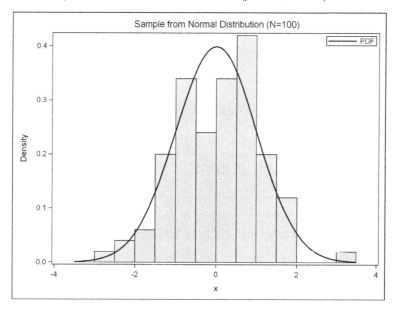

To explicitly specify values of the location and scale parameters, define `mu` and `sigma` outside of the DO loop, and then use the following statement inside the DO loop:

```
x = rand("Normal", mu, sigma);        /* X ~ N(mu, sigma) */
```

If X is a random variable from the $N(\mu, \sigma)$ distribution, then the expected value of X is μ and the variance is σ^2. Be aware that some authors denote the normal distribution by $N(\mu, \sigma^2)$, where the second parameter indicates the *variance*. This book uses $N(\mu, \sigma)$ instead, which matches the meaning of the parameters in the RAND function.

2.5.2   The Uniform Distribution

The uniform distribution is one of the most useful distributions in statistical simulation. One reason is that you can use the uniform distribution to sample from a finite set. Another reason is that "random variates with various distributions can be obtained by cleverly manipulating" independent, identically distributed uniform variates (Devroye 1986, p. 206).

The uniform distribution on the interval (a, b) is denoted by $U(a, b)$. Its density is given by $f(x) = (b - a)^{-1}$ for x in (a, b). The standardized uniform distribution on [0,1] (often called *the* uniform distribution) is denoted $U(0, 1)$.

You can use the following statement in the DATA step to generate a random observation from the standard uniform distribution:

```
x = rand("Uniform");              /* X ~ U(0, 1) */
```

The uniform random number generator never generates the number 0 nor the number 1. Therefore, all values are in the open interval $(0, 1)$.

You can also use the uniform distribution to sample random values from $U(a, b)$. To do this, define **a** and **b** outside of the DO loop, and then use the following statement inside the DO loop:

```
y = a + (b-a)*rand("Uniform");     /* Y ~ U(a, b) */
```

If X is a random variable from the standard uniform distribution, then the expected value of X is $1/2$ and the variance is $1/12$. In general, the uniform distribution on (a, b) has a uniform density of $1/(b - a)$. If Y is a random variable from the $U(a, b)$, the expected value of Y is $(a + b)/2$ and the variance is $(b - a)^2/12$.

Exercise 2.4: Generate 100 observations from a uniform distribution on the interval $(-1, 1)$.

2.5.3   The Exponential Distribution

The exponential distribution models the time between events that occur at a constant average rate. The exponential distribution is a continuous analog of the geometric distribution. The classic usage of the exponential distribution is to model the time between detecting particles emitted during radioactive decay.

The exponential distribution with scale parameter σ is denoted $\text{Exp}(\sigma)$. Its density is given by $f(x) = (1/\sigma) \exp(-x/\sigma)$ for $x > 0$. Alternatively, you can use $\lambda = 1/\sigma$, which is called the *rate parameter*. The rate parameter describes the rate at which an event occurs.

The following DATA step generates a random sample from the exponential distribution with scale parameter $\sigma = 10$. A histogram of the sample is shown in Figure 2.8.

```
data Exponential(keep=x);
call streaminit(4321);
sigma = 10;
do i = 1 to &N;
   x = sigma * rand("Exponential");     /* X ~ Expo(10) */
   output;
end;
run;
```

Figure 2.8 Sample from the Exponential Distribution ($\sigma = 10$) Overlaid with PDF

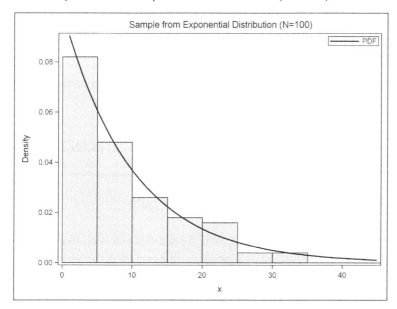

Notice that the scale parameter for the exponential distribution is not supported by the RAND function as of SAS 9.3. However, you can show that if X is distributed according to an exponential distribution with unit scale parameter, then $Y = \sigma X$ is distributed exponentially with scale parameter σ. The expected value of X is σ; the variance is σ^2. For example, the data shown in Figure 2.8 have a mean close to $\sigma = 10$.

If you use the exponential distribution with a scale parameter frequently, you might want to define and use the following SAS macro, which is used in Chapter 7 and in Chapter 12:

```
%macro RandExp(sigma);
   ((&sigma) * rand("Exponential"))
%mend;
```

The following statement shows how to call the macro from the DATA step:

```
x = %RandExp(sigma);
```

Exercise 2.5: Some distributions include the exponential distribution for particular values of the distribution parameters. For example, a Weibull(1, b) distribution is an exponential distribution with scale parameter b. Modify the program in this section to simulate data as follows:

```
x = rand("Weibull", 1, sigma);
```

Do you obtain a similar distribution of values? Use PROC UNIVARIATE to fit the exponential model to the simulated data.

2.6   Simulating Univariate Data in SAS/IML Software

You can also generate random samples by using the RANDGEN subroutine in SAS/IML software. The RANDGEN subroutine uses the same algorithms as the RAND function, but it fills an entire matrix at once, which means that you do not need a DO loop.

The following SAS/IML pseudocode simulates **N** observations from a named distribution:

```
%let N = 100;
proc iml;
call randseed(4321);                    /* or use a different seed   */
x = j(1, &N);                           /* allocate vector or matrix */
call randgen(x, "DistribName", param1, param2,...);     /* fill x */
```

The PROC IML program contains the following function calls:

1. A call to the RANDSEED subroutine, which specifies the seed that initializes the random number stream. If the argument is a positive integer, then the sequence is reproducible. Otherwise, the system time is used to initialize the random number stream, and the sequence will be different each time that you run the program.

2. A call to the J function, which allocates a matrix of a certain size. The syntax `J(r, c)` creates an $r \times c$ matrix. For this example, **x** is a vector that has one row and **N** columns.

3. A call to the RANDGEN subroutine, which fills the elements of **x** with random values from a named distribution. The supported distributions are listed in Section 2.7.

When you use the J function to allocate a SAS/IML matrix, the matrix is filled with 1s by default. However, you can use an optional third argument to fill the matrix with another value. For example `y=j(1,5,0)` allocates a 1×5 vector where each element has the value 0, and `y=j(4,3,.)` allocates a 4×3 matrix where each element is a SAS missing value.

Notice that the SAS/IML implementation is more compact than the DATA step implementation. It does not create a SAS data set, but instead holds the simulated data in memory in the **x** vector. By not writing a data set, the SAS/IML program is more efficient. However, both programs are blazingly fast. On the author's PC, generating a million observations with the DATA step takes about 0.2 seconds. Simulating the same data in PROC IML takes about 0.04 seconds.

2.6.1   Simulating Discrete Data

The RANDGEN subroutine in SAS/IML software supports the same distributions as the RAND function. Because the IML procedure does not need to create a data set that contains the simulated data, well-written simulations in the SAS/IML language have good performance characteristics.

The previous sections showed how to use the DATA step to generate random data from various distributions. The following SAS/IML program generates samples of size $N = 100$ from the same set of distributions:

```
proc iml;
/* define parameters */
p = 1/2;   lambda = 4;   k = 6;   prob = {0.5 0.2 0.3};

/* allocate vectors */
N = 100;
Bern = j(1, N);    Bino = j(1, N);    Geom = j(1, N);
Pois = j(1, N);    Unif = j(1, N);    Tabl = j(1, N);

/* fill vectors with random values */
call randseed(4321);
call randgen(Bern, "Bernoulli", p);      /* coin toss              */
call randgen(Bino, "Binomial", p, 10);   /* num heads in 10 tosses */
call randgen(Geom, "Geometric", p);      /* num trials until success */
call randgen(Pois, "Poisson", lambda);   /* num events per unit time */
call randgen(Unif, "Uniform");           /* uniform in (0,1)       */
Unif = ceil(k * Unif);                    /* roll die with k sides  */
call randgen(Tabl, "Table", prob);       /* sample with replacement */
```

Notice that in the SAS/IML language, which supports vectors in a natural way, the syntax for the "Table" distribution is simpler than in the DATA step. You simply define a vector of parameters and pass the vector to the RANDGEN subroutine. For example, you can use the following SAS/IML program to simulate data from a discrete uniform distribution as described in Section 2.4.4. The program simulates the roll of a six-sided die by using the RANDGEN subroutine to sample from six outcomes with equal probability:

```
proc iml;
call randseed(4321);
prob = j(6, 1, 1)/6;             /* equal prob. for six outcomes */
d = j(1, &N);                    /* allocate 1 x N vector        */
call randgen(d, "Table", prob);  /* fill with integers in 1-6    */
```

2.6.2 Sampling from Finite Sets

It can be useful to sample from a finite set of values. The SAS/IML language provides three functions that you can use to sample from finite sets:

- The RANPERM function generates random permutations of a set with n elements. Use this function to sample without replacement from a finite set with equal probability of selecting any item.

- The RANPERK function (introduced in SAS/IML 12.1) generates random permutations of k items that are chosen from a set with n elements. Use this function to sample k items without replacement from a finite set with equal probability of selecting any item.

- The SAMPLE function (introduced in SAS/IML 12.1) generates a random sample from a finite set. Use this function to sample with replacement or without replacement. This function can sample with equal probability or with unequal probability.

Each of these functions uses the same random number stream that is set by the RANDSEED routine. DATA step versions of the RANPERM and RANPERK functions are also supported.

These functions are similar to the "Table" distribution in that you can specify the probability of sampling each element in a finite set. However, the "Table" distribution only supports sampling with replacement, whereas these functions are suitable for sampling without replacement.

As an example, suppose that you have 10 socks in a drawer as in Section 2.4.5. Five socks are black, two socks are brown, and three socks are white. The following SAS/IML statements simulate three possible draws, without replacement, of five socks. The results are shown in Figure 2.9.

```
proc iml;
call randseed(4321);
socks = {"Black" "Black" "Black" "Black" "Black"
         "Brown" "Brown" "White" "White" "White"};
params = { 5,                          /* sample size             */
           3 };                        /* number of samples       */
s = sample(socks, params, "WOR");      /* sample without replacement */
print s;
```

Figure 2.9 Random Sample without Replacement

s				
White	Black	White	Black	Brown
Brown	Brown	Black	Black	Black
White	Black	White	Black	Brown

The SAMPLE function returns a 3×5 matrix, **s**. Each row of **s** is an independent draw of five socks (because **param[1] = 5**). After each draw, the socks are returned to the drawer and mixed well. The experiment is repeated three times (because **param[2] = 3**). Because each draw is without replacement, no row can have more than two brown socks or more than three white socks.

2.6.3   Simulating Continuous Data

Section 2.6.1 shows how to simulate data from discrete distributions in SAS/IML software. In the same way, you can simulate data from continuous distributions by calling the RANDGEN subroutine. As before, if you allocate a vector or matrix, then a single call of the RANDGEN subroutine fills the entire matrix with random values.

The following SAS/IML program generates samples of size $N = 100$ from the normal, uniform, and exponential distributions:

```
proc iml;
/* define parameters */
mu = 3;   sigma = 2;

/* allocate vectors */
N = 100;
StdNor = j(1, N);  Normal = j(1, N);
Unif = j(1, N);    Expo  = j(1, N);
```

```
/* fill vectors with random values */
call randseed(4321);
call randgen(StdNor,  "Normal");              /* N(0,1)       */
call randgen(Normal,  "Normal", mu, sigma);   /* N(mu,sigma)  */
call randgen(Unif,    "Uniform");             /* U(0,1)       */
call randgen(Expo,    "Exponential");         /* Exp(1)       */
```

2.7   Univariate Distributions Supported in SAS Software

In SAS software, the RAND function in Base SAS software and the RANDGEN subroutine in SAS/IML software are the main tools for simulating data from "named" distributions. These two functions call the same underlying numerical routines for computing random variates. However, there are some differences, as shown in Table 2.1:

Table 2.1   Differences Between RAND and RANDGEN Functions

	RAND Function	RANDGEN Subroutine
Called from:	DATA step	PROC IML
Seed set by:	CALL STREAMINT	CALL RANDSEED
Returns:	Scalar value	Vector or matrix of values

Because SAS/IML software can call Base SAS functions, it is possible to call the RAND function from a SAS/IML program. However, this is rarely done because it is more efficient to use the RANDGEN subroutine to generate many random variates with a single call.

Table 2.2 and Table 2.3 list the discrete and continuous distributions that are built into SAS software. Except for the t, F, and "NormalMix" distributions, you can identify a distribution by its first four letters. Parameters for each distribution are listed after the distribution name. Parameters in angled brackets are optional. If an optional parameter is omitted, then the default value is used.

The functions marked with an asterisk are supported by the RANDGEN function in SAS/IML 12.1. In general, parameters named μ and θ are location parameters, whereas σ denotes a scale parameter.

Table 2.2   Parameters for Discrete Distributions

Distribution	distname	parm1	parm2	parm3
Bernoulli	'BERNOULLI'	p		
Binomial	'BINOMIAL'	p	n	
Geometric	'GEOMETRIC'	p		
Hypergeometric	'HYPERGEOMETRIC'	N	R	n
Negative Binomial	'NEGBINOMIAL'	p	k	
Poisson	'POISSON'	m		
Table	'TABLE'	p		

Table 2.3 Parameters for Continuous Distributions

Distribution	distname	parm1	parm2	parm3
Beta	'BETA'	a	b	
Cauchy	'CAUCHY'			
Chi-Square	'CHISQUARE'	d		
Erlang	'ERLANG'	a	$<\sigma = 1>$	
Exponential	'EXPONENTIAL'	$<\sigma = 1>$		
F	'F'	n	d	
Gamma	'GAMMA'	a	$<\sigma = 1>$	
Laplace*	'LAPLACE'	$<\theta = 0>$	$<\sigma = 1>$	
Logistic*	'LOGISTIC'	$<\theta = 0>$	$<\sigma = 1>$	
Lognormal	'LOGNORMAL'	$<\mu = 0>$	$<\sigma = 1>$	
Normal	'NORMAL'	$<\mu = 0>$	$<\sigma = 1>$	
Normal Mixture*	'NORMALMIX'	p	μ	σ
Pareto*	'PARETO'	a	$<k = 1>$	
t	'T'	d		
Triangle	'TRIANGLE'	h		
Uniform	'UNIFORM'	$<a = 0>$	$<b = 1>$	
Wald*	'WALD' or 'IGAUSS'	λ	$<\mu = 1>$	
Weibull	'WEIBULL'	a	b	

Densities for all supported distributions are included in the documentation for the RAND function in *SAS Functions and CALL Routines: Reference*.

2.8 References

Devroye, L. (1986), *Non-uniform Random Variate Generation*, New York: Springer-Verlag. URL http://luc.devroye.org/rnbookindex.html

Wicklin, R. (2010), *Statistical Programming with SAS/IML Software*, Cary, NC: SAS Institute Inc.

Chapter 3
Preliminary and Background Information

Contents

3.1 What Every Programmer Should Know before Simulating Data

This chapter describes basic statistical distributional theory and details of SAS software that are essential to know before you begin simulating data. This chapter lays the foundation for how to efficiently simulate data in SAS software. In particular, this chapter describes the following:

- How to use four essential functions for working with statistical distributions. In SAS software, these functions are the PDF, CDF, QUANTILE, and RAND (or RANDGEN) functions.

- How to control random number streams in SAS software.

- How to check the correctness of simulated data.

- How to control output from SAS procedures by using the Output Delivery System (ODS).

If you are already familiar with these topics, then you can skip this chapter.

3.2 Essential Functions for Working with Statistical Distributions

If you work with statistical simulation, then it is essential to know how to compute densities, cumulative densities, quantiles, and random values for statistical distributions. In SAS software, the four functions are computed by using the PDF, CDF, QUANTILE, and RAND (or RANDGEN) functions.

- The PDF function evaluates the probability density function for continuous distributions and the probability mass function (PMF) for discrete distributions. For continuous distributions, the PDF function returns the probability density at the specified value. For discrete distributions, it returns the probability that a random variable takes on the specified value.

- The CDF function computes the cumulative distribution function. The CDF returns the probability that an observation from the specified distribution is less than or equal to a particular value. For continuous distributions, this is the area under the density curve up to a certain point.

- The QUANTILE function is closely related to the CDF function, but solves an inverse problem. Given a probability, P, the quantile for P is the smallest value, q, for which CDF(q) is greater than or equal to P

- The RAND function generates a random sample from a distribution. In SAS/IML software, use the RANDGEN subroutine, which fills up an entire matrix with random values. Examples of using these functions are given in Chapter 2, "Simulating Data from Common Univariate Distributions," and throughout this book.

3.2.1 The Probability Density Function (PDF)

The probability density function is often used to define a distribution. For example, the PDF for the standard normal distribution is $\psi(x) = (1/\sqrt{2\pi}) \exp(-x^2/2)$. You can use the PDF function to draw the graph of the probability density function. For example, the following SAS program uses the DATA step to generate points on the graph of the standard normal density, as follows:

```
data pdf;
do x = -3 to 3 by 0.1;
   y = pdf("Normal", x);
   output;
end;
run;
```

Figure 3.1 Standard Normal Probability Density

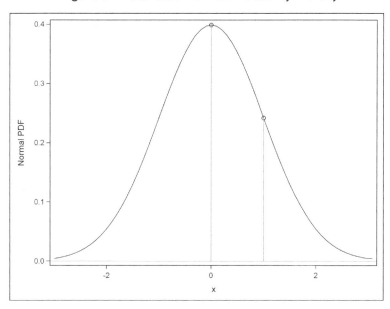

Figure 3.1 shows the graph of the standard normal PDF and the evaluation of the PDF at the points $x = 0$ and $x = 1$. The graph shows that $\phi(0)$ is a little less than 0.4 and that $\phi(1)$ is close to 0.25. This book's Web site provides SAS code that generates this and all other figures in this book.

Exercise 3.1: Use the SGPLOT procedure to plot the PDF of the standard exponential distribution.

3.2.2 The Cumulative Distribution Function (CDF)

The cumulative distribution function returns the probability that a value drawn from a given distribution is less than or equal to a given value.

For a continuous distribution, the CDF is the integral of the PDF from the lower range of the distribution (often $-\infty$) to the given value. For example, for the standard normal distribution, the CDF at $x = 0$ is 0.5 because that is the area under the normal curve to the left of 0. For a discrete distribution, the CDF is the sum of the PMF for all values less than or equal to the given value.

The CDF of any distribution is a non-decreasing function. For the familiar continuous distributions, the CDF is monotone increasing. For discrete distributions, the CDF is a step function. The following DATA step generates points on the graph of the standard normal CDF. Figure 3.2 shows the graph of the CDF for the standard normal distribution and the evaluation of the CDF at $x = 0$ and $x = 1.645$. The graph shows that CDF(0) = 0.5 and CDF(1.645) = 0.95.

```
data cdf;
do x = -3 to 3 by 0.1;
   y = cdf("Normal", x);
   output;
end;
run;
```

Figure 3.2 Standard Normal Cumulative Density

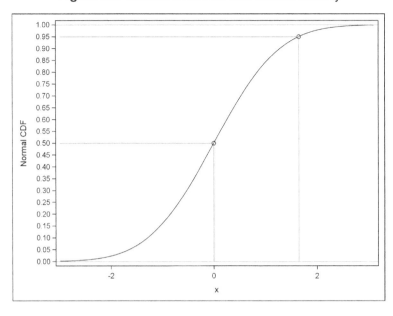

Exercise 3.2: Use the SGPLOT procedure to plot the CDF of the standard exponential distribution.

3.2.3   The Inverse CDF (QUANTILE Function)

If the CDF is continuous and strictly increasing, then there is a unique answer to the question: Given an area (probability), what is the value q for which the integral up to q has the specified area? The value q is called the quantile for the specified probability distribution. The median is the quantile of 0.5, the 90th percentile is the quantile of 0.9, and so on. For discrete distributions, the quantile is the smallest value for which the CDF is greater than or equal to the given probability.

For the standard normal distribution, the quantile of 0.5 is 0, and the 95th percentile is 1.645. You can find a quantile graphically by using the CDF plot: Choose a value q between 0 and 1 on the vertical axis, and use the CDF curve to find the value of x whose CDF is q, as shown in Figure 3.2.

Quantiles are used to compute p-values for hypothesis testing. For symmetric distributions, such as the normal distribution, the quantile for $1 - \alpha/2$ is used to compute two-sided p-values. For example, when $\alpha = 0.05$, the $(1 - \alpha/2)$ quantile for the standard normal distribution is computed by `quantile("Normal", 0.975)`, which is the famous number 1.96.

3.2.4   The Older SAS Random Number Functions

Long-time SAS users might remember the older PROB*XXX*, *XXX*INV, and RAN*XXX* functions for computing the CDF, inverse CDF (quantiles), and for sampling. For example, you can use the PROBGAM, GAMINV, and RANGAM functions for working with the gamma distribution. (For the normal distribution, the older names are PROBNORM, PROBIT, and RANNOR.) Although SAS still supports these older functions, you should not use them for simulations that generate many millions of random values. These functions use an older random number generator whose statistical properties are not as good as those of the newer Mersenne-Twister random number generator (Matsumoto and Nishimura 1998), which is known to have excellent statistical properties.

One of the desirable properties of the Mersenne-Twister algorithm is its extremely long period. The period of a random number stream is the number of uniform random numbers that you can generate before the sequence begins to repeat. The older RANXXX functions have a period of about 2^{31}, which is about 2×10^9, whereas the Mersenne-Twister algorithm has a period of about 2^{19937}, which is about 4×10^{6001}. For comparison, the number of atoms in the observable universe is approximately 10^{80}.

3.3   Random Number Streams in SAS

Often the term "random number" means a number sampled from the uniform distribution on $[0, 1]$. Random values for other distributions are generated by transformations of one or more uniform random variates. The seed value that controls the generation of uniform random variates also controls the generation of random variates for any distribution.

You can generate random numbers in SAS by using the RAND function in the DATA step or by using the RANDGEN subroutine in SAS/IML software. The functions call the same underlying numerical algorithms. These functions generate a stream of random variates from a specified distribution. You can control the stream by setting the seed for the random numbers. For the RAND function, the seed value is set by using the STREAMINIT subroutine; for the RANDGEN subroutine, use the RANDSEED subroutine in the SAS/IML language.

Specifying a particular seed enables you to generate the same set of random numbers every time that you run the program. This seems like an oxymoron: If the numbers are the same every time, then how can they be random? The resolution to this paradox is that the numbers that we call "random" should more accurately be called "pseudorandom" numbers. Pseudorandom numbers are generated by an algorithm, but have statistical properties of randomness. A good algorithm generates pseudorandom numbers that are indistinguishable from truly random numbers.

Why would you want a reproducible sequence of random numbers? Documentation and testing are two important reasons. Each example in this book explicitly sets a seed value (such as 4321) so that you can run the programs and obtain the same results.

3.3.1   Setting a Seed Value for Random Number Generation

The STREAMINIT subroutine is used to set the seed value for the RAND function in the DATA step. The seed value controls the sequence of random numbers. Syntactically, you should call the STREAMINIT subroutine one time per DATA step, prior to the first invocation of the RAND function. It does not make sense to call STREAMINIT multiple times within the same DATA step; subsequent calls are ignored.

A DATA step or SAS procedure produces the same pseudorandom numbers when given the same seed value. This means that you need to be careful when you write a single program that contains multiple DATA steps that each generate random numbers. In this case, use a different seed value in each DATA step or else the streams will not be independent. For example, the following DATA steps generate exactly the same simulated data:

```
data a;
call streaminit(4321);
do i = 1 to 10;  x=rand("uniform"); output;  end;
run;

data b;
call streaminit(4321);
do i = 1 to 10;  x=rand("uniform"); output;  end;
run;

proc compare base=a compare=b; run;       /* show they are identical */
```

This advice also applies to writing a macro function that generates random numbers. Do not explicitly set a seed value inside the macro. If you do, then each run of the macro will generate the same values. Rather, enable the user to specify the seed value in the syntax of the macro.

3.3.2   Setting a Seed Value from the System Time

If you do not explicitly call STREAMINIT, then the first call to the RAND function implicitly calls STREAMINIT with 0 as an argument. If you use a seed value of 0, then the STREAMINIT function uses the time on the internal system clock as the seed value. In this case, the random number stream is different each time you run the program.

However, "different each time" does not mean that the random number stream is not reproducible. SAS provides a macro variable named SYSRANDOM that contains the most recent seed value. If you explicitly set the seed value (for example, 4321), then the SYSRANDOM macro variable contains that value. If you specify a value of zero, then the system clock is used to generate a value such as 488637001. After the DATA step exits, the SYSRANDOM macro variable is set to that value. You can use the value as a seed to reproduce the same stream. For example, in the following program the first DATA step generates a random number stream by using the system clock. The SYSRANDOM macro is used in the second DATA step, which results in an identical set of random values, as shown in Figure 3.3.

```
data a;
call streaminit(0);                        /* different stream each time */
do i = 1 to 10;  x=rand("uniform"); output;  end;
run;

data b;
call streaminit(&sysrandom);       /* use SYSRANDOM to set same seed */
do i = 1 to 10;  x=rand("uniform"); output;  end;
run;

proc compare base=a compare=b short;       /* show they are identical */
run;
```

Figure 3.3 Identical Random Number Streams

```
The COMPARE Procedure
Comparison of WORK.A with WORK.B
(Method=EXACT)

NOTE: No unequal values were found. All values compared are exactly equal.
```

To show the seed value that was used, use the %PUT statement to display the value of the SYSRANDOM macro variable in the SAS log.

3.4 Checking the Correctness of Simulated Data

Whenever you implement an algorithm, it is a good idea to test whether your implementation is correct. Although it is usually quite difficult (or impossible) to rigorously test that simulated data are truly a sample from a specified distribution, there are four quick ways to check that you have not made a major error. Generate a large sample, and do one or more of the following:

- Compare sample moments of the simulated data with theoretical moments of the underlying distribution.

- Compare the empirical CDF with the theoretical cumulative distribution by using statistical goodness-of-fit tests.

- Compare the frequency or density of the simulated data with the theoretical PMF or PDF. To do this, create a bar chart or histogram of the simulated data and overlay the PMF or PDF, respectively, of the distribution.

- Compare empirical quantiles of the simulated data with theoretical quantiles. For continuous distributions, construct a quantile-quantile plot that compares quantiles of the simulated data to theoretical quantiles of the distribution. This is shown in Section 3.4.4. For discrete distributions, there are plots that serve a similar purpose (Friendly 2000).

3.4.1 Sample Moments and Goodness-of-Fit Tests

The UNIVARIATE procedure can be used to carry out these four checks for a variety of commonly used continuous distributions.

For example, the gamma distribution with shape parameter α and scale parameter σ is denoted by Gamma(α, σ). Its density is

$$f(x; \alpha, \sigma) = \frac{1}{\Gamma(\alpha)\sigma} \left(\frac{x}{\sigma}\right)^{\alpha-1} \exp\left(-\frac{x}{\sigma}\right), \quad x > 0$$

where $\Gamma(\alpha)$ is the gamma function. When α is an integer, $\Gamma(\alpha) = (\alpha - 1)!$. When $\sigma - 1$, the gamma distribution is denoted Gamma(α).

The following statements simulate observations from a gamma distribution with shape parameter $\alpha = 4$ and unit scale. PROC UNIVARIATE is then called to compute the sample moments (see Figure 3.4) and to perform a goodness-of-fit test (see Figure 3.5).

```
%let N=500;
data Gamma(keep=x);
call streaminit(4321);
do i = 1 to &N;
   x = rand("Gamma", 4);                      /* shape=4, unit scale */
   output;
end;
run;

/* fit Gamma distrib to data; compute GOF tests */
proc univariate data=Gamma;
   var x;
   histogram x / gamma(alpha=EST scale=1);
   ods select Moments ParameterEstimates GoodnessOfFit;
run;
```

Figure 3.4 Sample Moments

The UNIVARIATE Procedure
Variable: x

Moments			
N	500	Sum Weights	500
Mean	4.03734715	Sum Observations	2018.67357
Std Deviation	2.05277611	Variance	4.21388976
Skewness	0.98260471	Kurtosis	0.91340211
Uncorrected SS	10252.817	Corrected SS	2102.73099
Coeff Variation	50.8446768	Std Error Mean	0.09180294

For a gamma distribution with unit scale and shape parameter α, the mean and variance are α, the skewness is $2/\sqrt{\alpha}$, and the (excess) kurtosis is $6/\alpha$. When $\alpha = 4$, these values are 4, 4, 1, and 1.5. Figure 3.4 shows the sample moments for the simulated data, which are 4.04, 4.21, 0.98, and 0.91. The standard errors of the mean and variance are smaller than for the skewness and kurtosis, so you should expect the mean and variance to be closer to the corresponding population parameters than the skewness and kurtosis are.

Figure 3.5 shows the results of fitting a Gamma(α) distribution (with unit scale) to the simulated data. The maximum likelihood estimate of α is $\alpha = 4.03$. The goodness-of-fit tests all have large p-values, which indicates that there is no reason to doubt that the data are a random sample from a gamma distribution. However, as discussed in the documentation for the UNIVARIATE procedure in the *Base SAS Procedures Guide: Statistical Procedures*, "A test's ability to reject the null hypothesis (known as the *power* of the test) increases with the sample size." In practical terms, this means that for a large sample the goodness-of-fit tests often have small p-values, even if the sample is drawn from the fitted distribution.

Figure 3.5 Goodness-of-Fit Tests

The UNIVARIATE Procedure
Fitted Gamma Distribution for x

Parameters for Gamma Distribution		
Parameter	Symbol	Estimate
Threshold	Theta	0
Scale	Sigma	1
Shape	Alpha	4.03306
Mean		4.03306
Std Dev		2.008248

Goodness-of-Fit Tests for Gamma Distribution				
Test	Statistic		p Value	
Kolmogorov-Smirnov	D	0.03238942	Pr > D	>0.250
Cramer-von Mises	W-Sq	0.11322777	Pr > W-Sq	0.174
Anderson-Darling	A-Sq	0.64342001	Pr > A-Sq	>0.250

Exercise 3.3: Simulate $N = 10{,}000$ observations from the Gamma(4) distribution and compute the sample moments. Are the estimates closer to the parameter values than for the smaller sample size? What are the p-values for the goodness-of-fit tests?

3.4.2   Overlay a Theoretical PMF on a Frequency Plot

When you implement a sampling algorithm, it is good programming practice to check the validity of the algorithm. Overlaying the theoretical probabilities is one way to check whether the sample appears to be from the specified distribution. You can use the Graph Template Language (GTL) to define a template for a scatter plot overlaid on a bar chart. The bar chart displays the empirical frequencies of a sample whereas the scatter plot shows the probability mass function (PMF) for the population.

For example, suppose that you want to draw a random sample from the geometric distribution and overlay the PMF on the empirical frequencies. There are two common definitions for a geometric random variable. Let $B_1, B_2, \ldots$ be a sequence of independent Bernoulli trials with probability p. Then either of the following definitions are used for a geometric random variable with parameter p:

- the number of trials until the first success

- the number of trials up to but not including the first success

The two definitions are interchangeable because if X is a geometric random variable that satisfies the first definition, then $X - 1$ is a random variable that satisfies the second definition. SAS uses both definitions. The first definition is used by the RAND and RANDGEN functions, whereas (regrettably) the PDF function uses the second definition. Notice that a random variable that obeys the first definition takes on positive values; for the second definition, the variable is nonnegative.

The following DATA step draws 100 observations from the geometric distribution with parameter 0.5. Intuitively, this is a simulation of 100 experiments in which a coin is tossed until heads appears, and the number of tosses is recorded.

```
%let N=100;
data Geometric(keep=x);
call streaminit(4321);
do i = 1 to &N;
   x = rand("Geometric", 0.5);       /* number of tosses until heads */
   output;
end;
run;
```

The distribution of the simulated data can be compared with the PMF for the geometric distribution, which is computed by using the following DATA step:

```
/* For the geometric distribution, PDF("Geometric",t,0.5) computes the
   probability of t FAILURES, t=0,1,2,...  Use PDF("Geometric",t-1,0.5)
   to compute the number of TOSSES until heads appears, t=1,2,3,.... */
data PMF(keep=T Y);
do T = 1 to 9;
   Y = pdf("Geometric", T-1, 0.5);
   output;
end;
run;
```

The x variable in the Geometric data set contains 100 observations. The PMF data set contains nine observations. Nevertheless, you can merge these two data sets to create a single data set that contains all the data to be plotted:

```
data Discrete;
   merge Geometric PMF;
run;
```

The SGPLOT procedure does not support overlaying a bar chart and a scatter plot, so you need to use the GTL and PROC TEMPLATE to create a template that defines the layout of the graph. The template consists of three overlays: a bar chart, a scatter plot, and (optionally) a legend. The following statements define the template and render the graph by calling the SGRENDER procedure. Figure 3.6 shows the resulting graph. A short macro is used to handle the fact that the STAT= option in the BARCHART statement changed between SAS 9.3 and SAS 9.4.

```
/* GTL syntax changed at 9.4 */
%macro ScaleOpt;
   %if %sysevalf(&SysVer < 9.4) %then pct;   %else proportion;
%mend;
```

```
proc template;
define statgraph BarPMF;
dynamic _Title;                          /* specify title at run time */
begingraph;
   entrytitle _Title;
   layout overlay / yaxisopts=(griddisplay=on)
                    xaxisopts=(type=discrete);
   barchart     x=X / name='bar' legendlabel='Sample' stat=%ScaleOpt;
   scatterplot x=T y=Y / name='pmf' legendlabel='PMF';
   discretelegend 'bar' 'pmf';
   endlayout;
endgraph;
end;
run;

proc sgrender data=Discrete template=BarPMF;
   dynamic _Title = "Sample from Geometric(0.5) Distribution (N=&N)";
run;
```

Figure 3.6 Sample from Geometric Distribution ($p = 0.5$) Overlaid with PMF

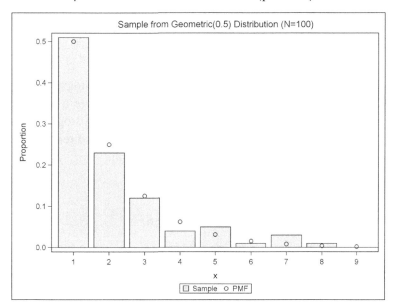

The BarPMF template is a simplified version of the template that is used for this book. (You can download the full version from the book's Web site.) Notice that the template uses a DYNAMIC statement to set the title of the graph. The title is set when the SGRENDER procedure is called. Although this is not necessary, it is convenient because you can reuse the template for many data sets. You can also use the DYNAMIC statement to specify other parts of the graph (for example, the names of the X, T, and Y variables) at run time.

Exercise 3.4: A negative binomial variable is defined as the number of failures before k successes in a series of independent Bernoulli trials with probability of success p. Sample 100 observations from a negative binomial distribution with $p = 0.5$ and $k = 3$ and overlay the PMF.

3.4.3   Overlaying a Theoretical Density on a Histogram

This section uses the Graph Template Language (GTL) to define a template for a density curve (PDF) overlaid on a histogram. For many distributions, you can use the HISTOGRAM statement in PROC UNIVARIATE to overlay a PDF on a histogram. However, using the GTL enables you to overlay density curves for families that are not supported by PROC UNIVARIATE.

As an example, suppose that you want to draw a random sample from the gamma distribution and overlay the PDF on the empirical distribution. Section 3.4.1 created the Gamma data set, which contains 500 observations from the Gamma(4) distribution. The PDF for the gamma distribution is computed by using the following DATA step:

```
data PDF(keep=T Y);
do T = 0 to 13 by 0.1;
   Y = pdf("Gamma", T, 4);
   output;
end;
run;
```

The Gamma data set contains 500 observations. The PDF data set contains fewer observations. You can merge these two data sets to create a single data set that contains all the data to be plotted:

```
data Cont;
   merge Gamma PDF;
run;
```

The SGPLOT procedure in SAS 9.3 does not support overlaying a histogram and a custom density curve, so you need to use the GTL and PROC TEMPLATE to create a template that defines the layout of the graph. The template consists of three overlays: a histogram, a series plot, and (optionally) a legend. The following statements define the template and render the graph by calling the SGRENDER procedure. Figure 3.7 shows the resulting graph.

```
proc template;
define statgraph HistPDF;
dynamic _Title _binstart _binstop _binwidth;
begingraph;
   entrytitle _Title;
   layout overlay / xaxisopts=(linearopts=(viewmax=_binstop));
   histogram X / scale=density endlabels=true xvalues=leftpoints
         binstart=_binstart binwidth=_binwidth;
   seriesplot x=T y=Y / name='PDF' legendlabel="PDF"
         lineattrs=(thickness=2);
   discretelegend 'PDF';
   endlayout;
endgraph;
end;
run;


proc sgrender data=Cont template=HistPDF;
dynamic _Title="Sample from Gamma(4) Distribution (N=&N)"
   _binstart=0                          /* left endpoint of first bin */
   _binstop=13                          /* right endpoint of last bin */
   _binwidth=1;                         /* width of bins              */
run;
```

Figure 3.7 Sample from Gamma(4) Distribution Overlaid with PDF

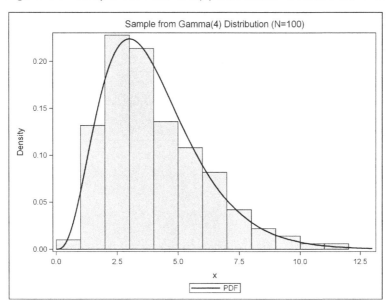

The HistPDF template is a simplified version of the template used for this book. (You can download the full version from the book's Web site.) Notice that the template uses a DYNAMIC statement to set the title of the graph and also to set three parameters that are used to specify the histogram bins. These parameters are set when the SGRENDER procedure is called. Specifying these parameters at run time enables you to use the template for many data sets. You can also define the names of the three variables as dynamic variables.

Exercise 3.5: Simulate 500 observations from a chi-square distribution with 3 degrees of freedom and overlay the PDF.

3.4.4   The Quantile-Quantile Plot

When you generate a sample from a continuous univariate distribution, you can use a quantile-quantile plot (Q-Q plot) to plot the ordered values of the sample against quantiles of the assumed distribution (Chambers et al. 1983). If the data are a sample from the specified distribution, then the points on the plot tend to fall along a straight line.

There are two ways to create a Q-Q plot in SAS. For many standard distributions, you can use the QQPLOT statement in PROC UNIVARIATE. In the following example, an exponential model is fit to simulated data. The Q-Q plot is shown in Figure 3.8.

```
data Exponential(keep=x);
call streaminit(4321);
sigma = 10;
do i = 1 to &N;
   x = sigma * rand("Exponential");
   output;
end;
run;
```

```
/* create an exponential Q-Q plot */
proc univariate data=Exponential;
   var x;
   qqplot x / exp;
run;
```

Figure 3.8 Exponential Q-Q Plot

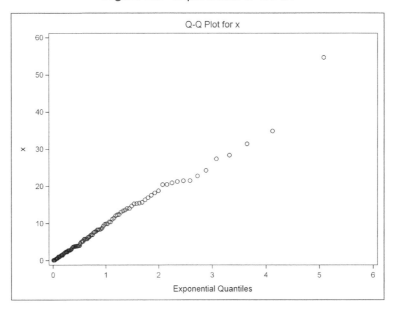

The points in the exponential Q-Q plot fall along a straight line. Therefore, there is no reason to doubt that the data are a random sample from an exponential distribution. The slope of the line is an estimate of the exponential scale parameter, which is 10 for this example. For more information about the interpretation of Q-Q plots, see "Interpretation of Quantile-Quantile and Probability Plots" in the PROC UNIVARIATE documentation in the *Base SAS Procedures Guide: Statistical Procedures*.

A second way to create a Q-Q plot is to manually compute the quantiles and use the SGPLOT procedure to create the graph. The advantage of this approach is that you can handle *any* distribution for which you can compute the inverse CDF. This approach requires the following steps:

1. Sort the data.

2. Compute *n* evenly spaced points in the interval (0,1), where *n* is the number of data points in the sample.

3. Compute the quantiles (inverse CDF) of the evenly spaced points.

4. Create a scatter plot of the sorted data versus the quantiles.

The steps are implemented by the following statements, which create a normal Q-Q plot. The Q-Q plot is shown in Figure 3.9.

```
%let N = 100;
data Normal(keep=x);
call streaminit(4321);
do i = 1 to &N;
   x = rand("Normal");                              /* N(0, 1) */
   output;
end;
run;

/* Manually create a Q-Q plot */
proc sort data=Normal out=QQ; by x; run;            /* 1 */

data QQ;
set QQ nobs=NObs;
v = (_N_ - 0.375) / (NObs + 0.25);                  /* 2 */
q = quantile("Normal", v);                          /* 3 */
label x = "Observed Data" q = "Normal Quantiles";
run;

proc sgplot data=QQ;                                /* 4 */
   scatter x=q y=x;
   xaxis grid;  yaxis grid;
run;
```

Figure 3.9 Normal Q-Q Plot

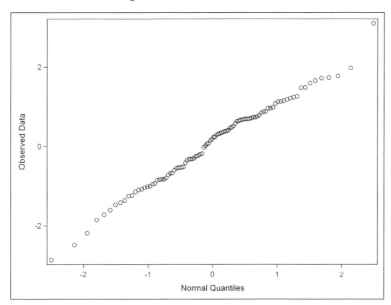

The points in the normal Q-Q plot fall along a straight line. Therefore, there is no reason to doubt that the simulated data are normally distributed. Furthermore, because the data in the Q-Q plot appear to pass through the origin and have unit slope, the parameter estimates for the normal fit to the data are close to $\mu = 0$ and $\sigma = 1$.

The previous technique enables you to compute a Q-Q plot for any continuous distribution. Simply replace the "Normal" parameter in the QUANTILE function with "Exponential" or "Lognormal" to obtain a Q-Q plot for those distributions.

Exercise 3.6: Simulate data from the $N(1,3)$ distribution. Create a normal Q-Q plot of the data. Do the points of the plot appear to fall along a line that has 1 as a y-intercept and 3 for a slope?

Exercise 3.7: Section 3.4 creates a Gamma data set that contains a random sample from the Gamma(4) distribution. Manually create a Q-Q plot, and verify that it is the same as is produced by the QQPLOT statement in PROC UNIVARIATE.

3.5   Using ODS Statements to Control Output

The SAS Output Delivery System (ODS) enables you to control the output from SAS programs. SAS procedures use ODS to display results. Most results are displayed as ODS tables; ODS statistical graphics are discussed briefly in Section 3.5.4.

There are ODS statements that enable you to control the output that is displayed, the destination for the output (such as the LISTING or HTML destinations), and many other aspects of the output. This section describes a few elementary ODS statements that are used in this book. For a more complete description, see the *SAS Output Delivery System: User's Guide*.

3.5.1   Finding the Names of ODS Tables

Before you can include, exclude, or save tables, you need to know the names of the tables. You can determine the names of ODS tables by using the ODS TRACE statement. The most basic syntax is as follows:

> **ODS TRACE ON | OFF ;**

When you turn on tracing, SAS software displays the names of subsequent ODS tables that are produced. The names are usually printed to the SAS log. For example, the following statements display the names of tables that are created by the FREQ procedure during a frequency analysis of the Sex variable in the Sashelp.Class data set:

```
ods trace on;
ods graphics off;
proc freq data=Sashelp.Class;
   tables sex / chisq;
run;
ods trace off;
```

The content of the SAS log is shown in Figure 3.10. The FREQ procedure creates two tables: the OneWayFreqs table and the OneWayChiSq table.

Figure 3.10 Names of ODS Tables

```
Output Added:
------------
Name:       OneWayFreqs
Label:      One-Way Frequencies
Template:   Base.Freq.OneWayFreqs
Path:       Freq.Table1.OneWayFreqs
------------

Output Added:
------------
Name:       OneWayChiSq
Label:      One-Way Chi-Square Test
Template:   Base.Freq.StatFactoid
Path:       Freq.Table1.OneWayChiSq
------------
NOTE: There were 19 observations read from the data set SASHELP.CLASS.
```

3.5.2 Selecting and Excluding ODS Tables

You can limit the output from a SAS procedure by using the ODS SELECT and ODS EXCLUDE statements. The ODS SELECT statement specifies the tables that you want to display; the ODS EXCLUDE statement specifies the tables that you want to suppress. The basic syntax is as follows:

ODS SELECT *table-names* | **ALL** | **NONE** ;

ODS EXCLUDE *table-names* | **ALL** | **NONE** ;

For example, PROC FREQ will display only the OneWayChiSq table if you specify either of the following statements:

```
ods select OneWayChiSq;
ods exclude OneWayFreqs;
```

3.5.3 Creating Data Sets from ODS Tables

You can use the ODS OUTPUT statement to create a SAS data set from an ODS table. For example, use the following statements to create a SAS data set named Freqs from the OneWayFreqs table that is produced by PROC FREQ:

```
proc freq data=Sashelp.Class;
   tables sex;
   ods output OneWayFreqs=Freqs;
run;
```

You can then use the data set as input for another procedure. This technique is especially useful when you need to save a statistic that appears in a table, but the procedure does not provide an option that writes the statistic to an output data set.

Of course, you typically need to know the names of the variables in the data set before you can use the data. You can use PROC CONTENTS to display the variable names as shown in the following statements:

```
proc contents data=Freqs short order=varnum;
run;
```

Figure 3.11 Variable Names

The CONTENTS Procedure

Variables in Creation Order
Table F_Sex Sex Frequency Percent CumFrequency CumPercent

3.5.4   Creating ODS Statistical Graphics

Many SAS procedures support ODS statistical graphics. You can use the ODS GRAPHICS statement to initialize the ODS statistical graphics system. The basic syntax is as follows:

ODS GRAPHICS ON | OFF ;

You can use the ODS SELECT and ODS EXCLUDE statements on graphical objects just as you can for tabular output. For example, the following statements perform a frequency analysis on the Age variable in the Sashelp.Class data set. The ODS SELECT statement excludes all output except for the bar chart that is shown in Figure 3.12.

```
ods graphics on;
proc freq data=Sashelp.Class;
   tables age / plot=FreqPlot;
   ods select FreqPlot;
run;
```

Figure 3.12 An ODS Graphic

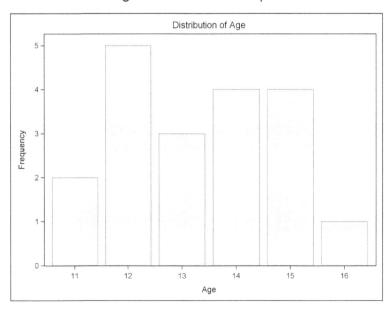

3.6   References

Chambers, J. M., Cleveland, W. S., Kleiner, B., and Tukey, P. A. (1983), *Graphical Methods for Data Analysis*, Belmont, CA: Wadsworth International Group.

Friendly, M. (2000), *Visualizing Categorical Data*, Cary, NC: SAS Institute Inc.

Matsumoto, M. and Nishimura, T. (1998), "Mersenne Twister: A 623-Dimensionally Equidistributed Uniform Pseudo-random Number Generator," *ACM Transactions on Modeling and Computer Simulation*, 8, 3–30.

Part II

Basic Simulation Techniques

Chapter 4

Simulating Data to Estimate Sampling Distributions

Contents

4.1 The Sampling Distribution of a Statistic

Chapter 2, "Simulating Data from Common Univariate Distributions," shows how to simulate a random sample with N observations from a variety of standard distributions. This chapter describes how to simulate *many* random samples and use them to estimate the distribution of sample statistics. You can use the sampling distribution to compute p-values, standard errors, and confidence intervals, just to name a few applications.

In data simulation, you generate a sample from a population model. If you compute a statistic for the simulated sample, then you get an estimate of a population quantity. For example, Figure 4.1 shows a random sample of size 50 drawn from some population. The third quartile of the sample is shown as a vertical line. It is an estimate for the third quartile of the population.

Figure 4.1 Random Sample and Quartile Statistic

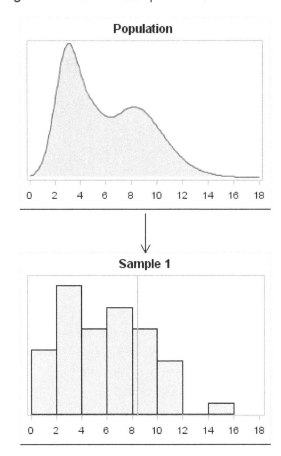

You can repeat this process by drawing another random sample of size 50 and computing another sample quartile. If you continue to repeat this process, then you will generate a slew of statistics, one for each random sample. In Figure 4.2, one of the samples has the quartile value 8.4, another sample has the quartile value 9.6, a third sample has the quartile value 6.9, and so on.

The union of the statistics is an *approximate sampling distribution* (ASD) for the statistic. This ASD is shown in the bottom graph of Figure 4.2, which shows a histogram for 1,000 statistics, each statistic being the third quartile for a random sample drawn from the population. The "width" of the ASD (for example, the standard deviation) indicates how much the statistic is likely to vary among random samples of the same size. In other words, the width indicates how much uncertainty to expect in an estimate due to sampling variability. In the figure, the standard deviation of the ASD is 0.7 and the interquartile range is 0.9.

The simulated sampling distribution is approximate because it is based on a finite number of samples. The exact sampling distribution requires computing the statistic for all possible samples of the given sample size. The shape of the sampling distribution (both exact and approximate) depends on the distribution of the population, the statistic itself, and the sample size. For some statistics, such as the sample mean, the theoretical sampling distribution is known. However, for many statistics the sampling distribution is known only by using simulation studies.

Figure 4.2 Accumulation of Many Statistics from Samples

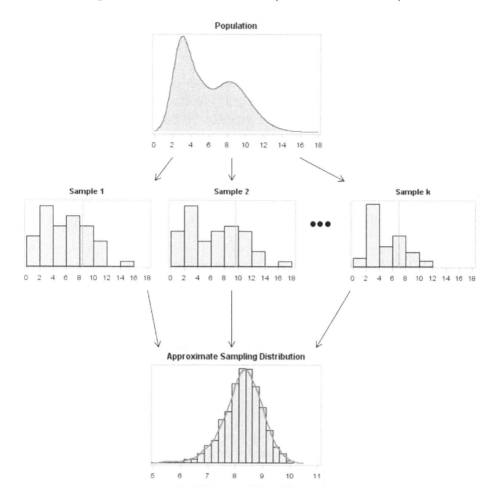

The sampling distribution of a statistic determines a familiar statistic: the standard error. The *standard error* of a statistic is the standard deviation of its sampling distribution.

The sampling distribution is used to estimate uncertainty in estimates. It is also useful for hypothesis testing. Suppose that a colleague collects 50 observations and tells you that the third quartile of the data is 6, and asks whether you think it is likely that her data are from the population in Figure 4.2. From the figure, you notice that the statistic for her sample is smaller than all but a few of the simulated sample quartiles. Therefore, you respond that it is unlikely that her sample came from the population. This demonstrates the main idea of hypothesis testing: You look at some sampling distribution and compute the probability of a random value being at least as extreme as the value that you are testing. Hypothesis testing and p-values are discussed in Chapter 5, "Using Simulation to Evaluate Statistical Techniques."

In summary, you can use simulation to construct an ASD for a statistic. The ASD enables you to compute quantities that are of statistical interest, such as the standard error and p-values for hypothesis tests.

4.2   Monte Carlo Estimates

The previous section describes the essential ideas of statistical simulation without using mathematics. This section uses statistical theory to describe the essential ideas. The presentation and notation are based on Ross (2006, p. 117–126), which is an excellent reference for readers who are interested in a more rigorous treatment of simulation than is presented in this book.

One use of simulation is to estimate some unknown quantity, θ, in the population by some statistic, X, whose expected value is θ. A statistic is a random variable. If you draw one sample from the population and compute the statistic, X_1, on that sample, then you have one estimate of θ. A second independent sample results in X_2, which is also an estimate of θ. After a large number (m) of samples, the union of the statistics is an approximation to the sampling distribution of X, given the population.

You can use the ASD of X to understand how the statistic can vary due to sampling variation. However, you can also use the ASD to form an improved estimate of θ. After m samples, you can compute the *Monte Carlo estimate* to θ, which is simply the average of the statistics: $\bar{X} = \Sigma_i^m X_i / m$. The Monte Carlo estimate (like the X_i statistics) is an estimate of θ. However, it has a smaller variance than the X_i. The X_i have a common variance, σ^2, because they are random variables from the same population. To determine how close the Monte Carlo estimate is to θ, you can compute the *mean square error*, which is the expected value of $(\bar{X} - \theta)^2$. The mean square error equals σ^2/m (Ross 2006, p. 118). The square root of this quantity, $\sigma/\sqrt{m}$, is a measure of how far the Monte Carlo estimate is from θ.

The Monte Carlo estimate, $\bar{X}$, is a good estimator of θ when $\sigma/\sqrt{m}$ is small. Of course, the population variance σ^2 is unknown, but is usually estimated by the sample variance $S^2 = \Sigma_{i=1}^m (X_i - \bar{X})^2/(m-1)$. The sample standard deviation $S = \sqrt{S^2}$ is used to estimate σ.

You can use the Monte Carlo estimate to estimate discrete probabilities and proportions. An example is given in Section 4.4.3.

4.3   Approximating a Sampling Distribution

A statistical simulation consists of the following steps:

1. Simulate many samples, each from the same population distribution.

2. For each sample, compute the quantity of interest. The quantity might be a count, a mean, or some other statistic. The union of the statistics is an ASD for the statistic.

3. Analyze the ASD to draw a conclusion. The conclusion is often an estimate of something, such as a standard error, a confidence interval, a probability, and so on.

This chapter describes how to carry out this scheme efficiently in SAS software. This chapter focuses on simulating data from a continuous distribution and computing the ASD for basic statistics such as means, medians, quantiles, and correlation coefficients.

There are two main techniques for implementing step 1 and step 2: the *BY-group technique* and the *in-memory technique*.

BY-group technique: Use this technique when you plan to use a SAS procedure to analyze each sample in step 2. In this approach, you write all samples to the same data set. Each sample is identified by a unique value of a variable, which is called `SampleID` in this book. Use a BY statement in the SAS procedure to efficiently analyze each sample.

In-memory technique: Use this technique when you plan to use the SAS/IML language to analyze each sample in step 2.

- If you are analyzing univariate data, then it is common to store each simulated sample as a row (or column) of a matrix. You can then compute a statistic on each row (or column) to estimate the sampling distribution.

- If you are analyzing multivariate data, then it is common to write a DO loop. At each iteration, simulate data, compute the statistic on that data, and store the statistic in a vector. When the DO loop completes, the vector contains the ASD.

It is also possible to combine the techniques: If the simulated data are in a SAS data set but you intend to analyze the data in the SAS/IML language, you can read each BY group (sample) into a SAS/IML matrix or vector by using a DO loop and a WHERE clause in the READ statement.

4.4 Simulation by Using the DATA Step and SAS Procedures

This section provides several examples of using the DATA step to simulate data. The resulting data are analyzed by using SAS procedures, which compute statistics on the data by using the BY statement to iterate over all samples. The ASD of the statistic is analyzed by using a procedure such as PROC MEANS or PROC UNIVARIATE.

Each simulation in this chapter uses a fixed seed value so that you can reproduce the analysis and graphs. If you use a different value, then the graphs and analyses will change slightly, but the main ideas remain the same.

4.4.1 A Template for Simulation with the DATA Step and Procedures

To generate multiple samples with the DATA step, use the techniques that are described in Chapter 2, but put an extra DO loop around the statements that generate each sample. This outer loop is the *simulation loop*, and the looping variable identifies each sample. For example, suppose that you want to generate m samples, and each sample contains N observations. You can use the following steps to generate and analyze the simulated data:

1. Simulate data with known properties. The resulting data set contains $N \times m$ observations and a variable SampleID that identifies each sample. (Data that are arranged like this are said to be in the "long" format.) The SampleID variable must be in sorted order so that you can use it as

a BY-group variable. If you generate the data in a different order, then sort the data by using the SORT procedure.

2. Use some procedure and the BY statement to compute statistics for each sample. Store the *m* results in an output data set.

3. Analyze the statistics in the output data set. Compute summary statistics and graph the approximate sampling distribution to answer questions about confidence intervals, hypothesis tests, and so on.

The following pseudocode is a template for these steps:

```
%let N = 10;                          /* size of each sample */
%let NumSamples = 100;                /* number of samples   */
/* 1. Simulate data with DATA step */
data Sim;
do SampleID = 1 to &NumSamples;
   do i = 1 to &N;
      x = rand("DistribName", param1, param2, ...);
      output;
   end;
end;
run;


/* 2. Compute statistics for each sample */
proc <NameOfProc> data=Sim out=OutStats noprint;
   by SampleID;
   /* compute statistics for each sample */
run;


/* 3. Analyze approx. sampling distribution of statistic */
proc univariate data=OutStats;       /* or PROC MEANS or PROC CORR   */
   /* analyze sampling distribution */
run;
```

The Sim data set contains the SampleID variable, which has the values 1–100. The x variable contains random values from the specified distribution. The first 10 observations are the first simulated sample, the next 10 are the second sample, and so forth.

The second step (computing statistics and storing them in the OutStats data set) varies a little, depending on the procedure that you use to compute the statistics. There are four common ways to create the OutStats data set from a SAS procedure:

• By using an option in the PROC statement. For example, the OUTEST= option in the PROC REG statement creates an output data set that contains parameter estimates.

• By using an OUTPUT statement. For example, the MEANS procedure has an OUTPUT statement that writes univariate statistics to an output data set.

• By using an OUT= option in a procedure statement. For example, the TABLES statement in PROC FREQ has an OUT= option that writes frequencies and expected frequencies to an output data set.

• By using an ODS OUTPUT statement to write an ODS table to a data set.

The following sections apply this general template to specific examples. See Section 3.5 if you are not comfortable using ODS to select, exclude, and output tables and graphics.

4.4.2   The Sampling Distribution of the Mean

You can use a simple simulation to approximate the sampling distribution of the mean, which leads to an estimate for the standard error. This is a good example to start with because the sampling distribution of the mean is well known due to the *central limit theorem* (CLT). The CLT states that the sampling distribution of the mean is approximately normally distributed, with a mean equal to the population mean, μ, and a standard deviation equal to $\sigma/\sqrt{N}$, where σ is the standard deviation of the population (Ross 2006). This approximation gets better as the sample size, N, gets larger.

What is interesting about the CLT is that the result holds regardless of the distribution of the underlying population.

The following SAS statements generate 1,000 samples of size 10 that are drawn from a uniform distribution on the interval [0, 1]. The MEANS procedure computes the mean of each sample and saves the means in the OutStatsUni data set.

```
%let N = 10;                            /* size of each sample */
%let NumSamples = 1000;                 /* number of samples   */
/* 1. Simulate data */
data SimUni;
call streaminit(123);
do SampleID = 1 to &NumSamples;
   do i = 1 to &N;
      x = rand("Uniform");
      output;
   end;
end;
run;


/* 2. Compute mean for each sample */
proc means data=SimUni noprint;
   by SampleID;
   var x;
   output out=OutStatsUni mean=SampleMean;
run;
```

The simulation generates the ASD of the sample mean for $U(0, 1)$ data with $N = 10$. The simulated sample means are contained in the variable SampleMean in the OutStatsUni data set. You can summarize the ASD by using the MEANS procedure (see Figure 4.3), and visualize it by using the UNIVARIATE procedure (see Figure 4.4), as follows:

```
/* 3. Analyze ASD: summarize and create histogram */
proc means data=OutStatsUni N Mean Std P5 P95;
   var SampleMean;
run;
```

```
ods graphics on;                                /* use ODS graphics   */
proc univariate data=OutStatsUni;
   label SampleMean = "Sample Mean of U(0,1) Data";
   histogram SampleMean / normal;               /* overlay normal fit */
   ods select Histogram;
run;
```

Figure 4.3 Summary of the Sampling Distribution of the Mean of $U(0, 1)$ Data, $N = 10$

The MEANS Procedure

Analysis Variable : SampleMean				
N	Mean	Std Dev	5th Pctl	95th Pctl
1000	0.5026407	0.0925483	0.3540121	0.6588903

Figure 4.4 Approximate Sampling Distribution of the Sample Mean of $U(0, 1)$ Data, $N = 10$

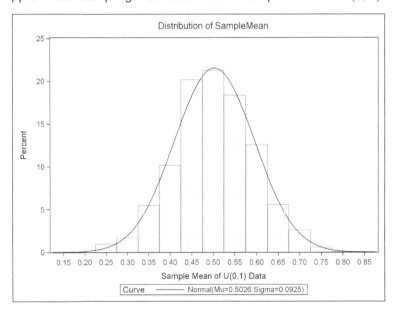

Figure 4.3 shows descriptive statistics for the SampleMean variable. The Monte Carlo estimate of the mean is 0.5, and the standard error of the mean is 0.09. Figure 4.4 shows a histogram of the SampleMean variable, which appears to be approximately normally distributed.

You can use statistical theory to check whether the simulated values make sense. The $U(0, 1)$ distribution has a population mean of $1/2$ and a population variance of $\sigma^2 = 1/12$. The CLT states that the sampling distribution of the mean is approximately normally distributed with mean $1/2$ and standard deviation $\sigma/\sqrt{N}$ for large values of N. Even though N is small for this example, the standard deviation is nevertheless close to $1/\sqrt{12N} \approx 0.0913$.

The P5 and P95 options in the PROC MEANS statement make it easy to compute a 90% confidence interval for the mean. To compute a 95% confidence interval, use PROC UNIVARIATE to compute the 2.5 and 97.5 percentiles of the data, as follows:

```
proc univariate data=OutStatsUni noprint;
   var SampleMean;
   output out=Pctl95 N=N mean=Mean pctlpts=2.5 97.5 pctlpre=Pctl;
run;

proc print data=Pctl95 noobs;
run;
```

Figure 4.5 95% Confidence Intervals

N	Mean	Pctl2_5	Pctl97_5
1000	0.50264	0.31462	0.68397

Exercise 4.1: The standard exponential distribution has a mean and variance that are both 1. Generate 1,000 samples of size $N = 64$, and compute the mean and standard deviation of the ASD. Is the standard deviation close to $1/8$?

4.4.3 Using a Sampling Distribution to Estimate Probability

Section 4.4.2 simulates samples of size $N = 10$ from a $U(0, 1)$ distribution. Suppose that someone draws a new, unseen, sample of size 10 from $U(0, 1)$. You can use the simulated results shown in Figure 4.4 to answer the following question: What is the probability that the mean of the sample is greater than 0.7?

You can write a short DATA step to compute the proportion of simulated means that satisfy this condition, as follows:

```
data Prob;
   set OutStatsUni;
   LargeMean = (SampleMean>0.7);      /* create indicator variable */
run;

proc freq data=Prob;
   tables LargeMean / nocum;          /* compute proportion       */
run;
```

Figure 4.6 Estimated Probability of a Mean Greater Than 0.7

The FREQ Procedure

LargeMean	Frequency	Percent
0	981	98.10
1	19	1.90

The answer is that only 1.9% of the simulated means are greater than 0.7. It is therefore rare to observe a mean this large in a random draw of size 10 from the uniform distribution.

Exercise 4.2: Instead of using the DATA step to create the LargeMean indicator variable, you can use the FORMAT procedure to define a SAS format that displays values that are less than 0.7 differently than values that are greater than 0.7, as follows:

```
proc format;
   value CutVal low-<0.7="less than 0.7"  0.7-high="greater than 0.7";
run;
```

Use the CUTVAL. format to estimate the probability that SampleMean is greater than 0.7.

4.4.4   The Sampling Distribution of Statistics for Normal Data

This section generates data from the standard normal distribution and examines the sampling distributions of several statistics: the mean, the median, and the variance. Because the underlying distribution is normal, the following results are known (Kendall and Stuart 1977):

- Given a sample with an odd number of points, for example $N = 2k + 1$, the variance of the sample mean is about 64% smaller than the variance of the sample median. More precisely, the ratio of the variances is $(4k)/(\pi(2k + 1))$, which approaches $2/\pi \approx 0.64$ for large k.

- The sampling distribution of the variance follows a scaled chi-square distribution with $N - 1$ degrees of freedom.

This section shows you how to use statistical simulation to demonstrate these facts and, by extension, other "textbook facts" about the sampling distribution of statistics. The value of this section is that you can use these same techniques when no theoretical result is available.

When you use simulation to demonstrate a statistical fact about a sampling distribution, you need to simulate a large number of samples so that the ASD is a good approximation to the exact sampling distribution. The following statements generate 10,000 random samples from $N(0, 1)$, each of size $N = 31$:

```
%let N = 31;                            /* size of each sample */
%let NumSamples = 10000;                /* number of samples   */
/* 1. Simulate data */
data SimNormal;
call streaminit(123);
do SampleID = 1 to &NumSamples;
   do i = 1 to &N;
      x = rand("Normal");
      output;
   end;
end;
run;


/* 2. Compute statistics for each sample */
proc means data=SimNormal noprint;
   by SampleID;
   var x;
   output out=OutStatsNorm mean=SampleMean median=SampleMedian var=SampleVar;
run;
```

The program generates three sampling distributions: one for the mean, one for the median, and one for the variance. All three sampling distributions are stored in the OutStatsNorm data set.

4.4.4.1 Variance of the Sample Mean and Median

Each simulated sample has $N = 31$ observations. Let $k = 15$ so that $N = 2k + 1$. For these simulated data, is it true that the variance of the sample mean is $(4k)/(\pi(2k + 1)) = 0.616$ times the variance of the sample median? You can use PROC MEANS to compute the variances, which are shown in Figure 4.7.

```
/* variances of sampling distribution for mean and median */
proc means data=OutStatsNorm Var;
   var SampleMean SampleMedian;
run;
```

Figure 4.7 Variances of the Sample Mean and Median of $N(0, 1)$ Data

The MEANS Procedure

Variable	Variance
SampleMean	0.0316398
SampleMedian	0.0490275

The simulation demonstrates that the theory works well. The ratio of the variances for the ASD in Figure 4.7 is 0.645, which is within 5% of the theoretical value.

It is interesting to overlay the two distributions to see how they compare. A simple way to compare two distributions is to plot their densities, as follows:

```
proc sgplot data=OutStatsNorm;
   title "Sampling Distributions of Mean and Median for N(0,1) Data";
   density SampleMean /  type=kernel legendlabel="Mean";
   density SampleMedian / type=kernel legendlabel="Median";
   refline 0 / axis=x;
run;
```

Figure 4.8 shows that the two distributions are approximately symmetric and have similar centers. However, the distribution of the median is wider, as predicted by theory.

Figure 4.8 Approximate Sampling Distribution for the Mean and Median of $N(0, 1)$ Data, $N = 31$

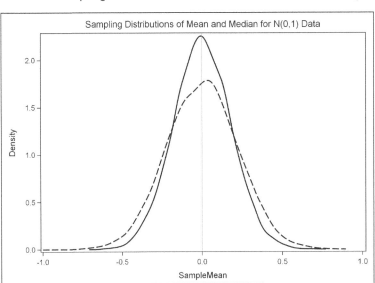

4.4.4.2   Sampling Distribution of the Variance

Statistical theory says that the quantity $(N - 1)s^2/\sigma^2$ follows a chi-square distribution with $N - 1$ degrees of freedom, where s^2 is the sample variance and σ^2 is the variance of the underlying normal population. For the simulated data in this example, $N = 31$ and $\sigma = 1$.

You can use PROC UNIVARIATE to fit a chi-square distribution to the scaled sample variances. Fitting a parametric distribution to data is accomplished with the HISTOGRAM statement. A chi-square distribution with d degrees of freedom is equivalent to a Gamma$(d/2, 2)$ distribution. Therefore, the following statements fit a chi-square distribution to the scaled variances. The resulting histogram is shown in Figure 4.9.

```
/* scale the sample variances by (N-1)/sigma^2 */
data OutStatsNorm;
   set OutStatsNorm;
   ScaledVar = SampleVar * (&N-1)/1;
run;

/* Fit chi-square distribution to data */
proc univariate data=OutStatsNorm;
   label ScaledVar = "Variance of Normal Data (Scaled)";
   histogram ScaledVar / gamma(alpha=15 sigma=2);   /* - chi-square */
   ods select Histogram;
run;
```

Figure 4.9 Scaled Distribution of Variance and Chi-Square Fit

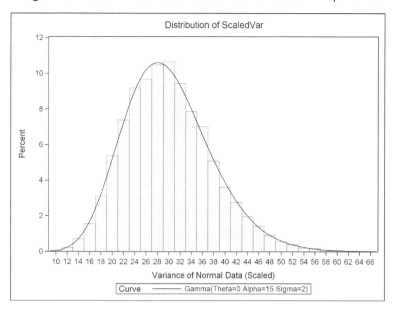

The goodness-of-fit statistics are not shown, but Figure 4.9 indicates that a Gamma(15, 2) distribution fits the sampling distribution. This is equivalent to a chi-square distribution with 30 degrees of freedom. This simulation agrees with the theoretical result that the variance of normal data follows a chi-square distribution.

4.4.5   The Effect of Sample Size on the Sampling Distribution

Section 4.4.2 demonstrates how to approximate the sampling distribution of the sample mean for samples of size 10 that are drawn from a uniform distribution. Statistical theory states that the sampling distribution of the sample mean is approximately normally distributed with mean $1/2$ and standard deviation $1/\sqrt{12N}$.

You can use simulation to illustrate the effect of sample size on the standard deviation of the sampling distribution. The following simulation draws 1,000 samples of size 10, 30, 50, and 100, and displays the Monte Carlo estimates in Figure 4.10:

```
%let NumSamples = 1000;              /* number of samples */
/* 1. Simulate data */
data SimUniSize;
call streaminit(123);
do N = 10, 30, 50, 100;
   do SampleID = 1 to &NumSamples;
      do i = 1 to N;
         x = rand("Uniform");
         output;
      end;
   end;
end;
run;
```

```
/* 2. Compute mean for each sample */
proc means data=SimUniSize noprint;
   by N SampleID;
   var x;
   output out=OutStats mean=SampleMean;
run;

/* 3. Summarize approx. sampling distribution of statistic */
proc means data=OutStats Mean Std;
   class N;
   var SampleMean;
run;
```

Figure 4.10 Means and Standard Deviations of Sampling Distribution of the Mean

The MEANS Procedure

		Analysis Variable : SampleMean	
N	N Obs	Mean	Std Dev
10	1000	0.5026407	0.0925483
30	1000	0.5009984	0.0544078
50	1000	0.4986503	0.0405417
100	1000	0.4993238	0.0286986

Notice that the simulation now includes the sample size, N, as a parameter that is varied during the simulation. There are actually four simulations being run, one for each sample size. The SAS syntax makes it easy to do the following:

- Simulate all of the data in a single DATA step by writing an extra DO loop outside of the usual simulation statements.

- Analyze all of the data by using both N and SampleID variables in the BY statement of the MEANS procedure.

- Display the mean and standard deviation of all the simulations by using the N variable in the CLASS statement in the second call to the MEANS procedure.

You can use the computed means and standard deviations to visualize the sampling distribution of the mean as the sample size increases. Figure 4.11 shows normal density distributions for the estimated parameters (μ_i, σ_i), $i = 1, \ldots, 4$.

Figure 4.11 Approximate Sampling Distribution of the Mean, $N = 10, 30, 50, 100$

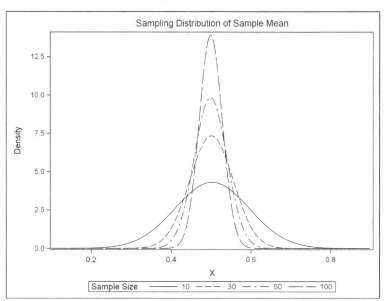

Exercise 4.3: Create Figure 4.11 by overlaying the normal PDFs whose parameters are given in Figure 4.10.

Exercise 4.4: The last column of Figure 4.10 shows that the standard error of the mean decreases as the sample size increases. Let the sample size, N, vary from 10 to 200 in increments of 10. Simulate 1,000 samples of size N from the uniform distribution. Plot the standard error of the mean as a function of the sample size.

4.4.6 Bias of Kurtosis Estimates in Small Samples

Although Section 4.4.1 shows how to write a DATA step that simulates data from a single distribution, you can easily extend the idea. This section uses a single DATA step to simulate data from four different distributions.

For small sample sizes drawn from distribution with large kurtosis, the sample kurtosis is known to underestimate the kurtosis of the population (Joanes and Gill 1998; Bai and Ng 2005). (Here "kurtosis" refers to "excess kurtosis"; see the discussion in Section 16.4.) In fact, Bai and Ng (2005, p. 49) comment:

> Measuring the tails [of a distribution by] using the kurtosis statistic is not a sound approach. ... The true value of [the kurtosis] will likely be substantially underestimated in practice, because a very large number of observations is required to get a reasonable estimate .... Exceptions are distributions with thin tails, such as the normal distribution.

Suppose that you want to run a simulation study to investigate the truth of the previous statement. You might choose a few distributions with known kurtosis, generate many small samples, compute the sample kurtosis for each sample, and compare the sample kurtosis to the known population value.

Table 4.1 shows the skewness and kurtosis for four distributions. The values for the lognormal distribution are approximate; the other values are exact.

Table 4.1   Skewness and Excess Kurtosis for Four Distributions

Distribution	Skewness	Kurtosis
Normal	0	0
t_5	0	6
Exponential	2	6
Lognormal(0, 0.503)	1.764	6

The following statements simulate 1,000 samples (each of size $N = 50$) from each distribution. PROC MEANS is used to compute the sample kurtosis and to save the values to the Moments data set.

```
/* bias of kurtosis in small samples */
%let N = 50;                          /* size of each sample */
%let NumSamples = 1000;               /* number of samples   */
data SimSK(drop=i);
call streaminit(123);
do SampleID = 1 to &NumSamples;       /* simulation loop              */
   do i = 1 to &N;                    /* N obs in each sample         */
      Normal      = rand("Normal");   /* kurt=0                       */
      t           = rand("t", 5);     /* kurt=6 for t, exp, and logn */
      Exponential = rand("Expo");
      LogNormal   = exp(rand("Normal", 0, 0.503));
      output;
   end;
end;
run;


proc means data=SimSK noprint;
   by SampleID;
   var Normal t Exponential LogNormal;
   output out=Moments(drop=_type_ _freq_) Kurtosis=;
run;
```

The KURTOSIS= option in the OUTPUT statement of PROC MEANS is used so that the kurtosis values in the Moments data set have the same names as the original variables. The following statements use PROC TRANSPOSE to transpose the data and PROC SGPLOT to create box plots of the four distributions. See Figure 4.12.

```
proc transpose data=Moments out=Long(rename=(col1=Kurtosis));
   by SampleID;
run;

proc sgplot data=Long;
   title "Kurtosis Bias in Small Samples: N=&N";
   label _Name_ = "Distribution";
   vbox Kurtosis / category=_Name_ meanattrs=(symbol=Diamond);
   refline 0 6 / axis=y;
   yaxis max=30;
   xaxis discreteorder=DATA;
run;
```

Figure 4.12 Distribution of Sample Kurtosis for Small Samples, $N = 50$

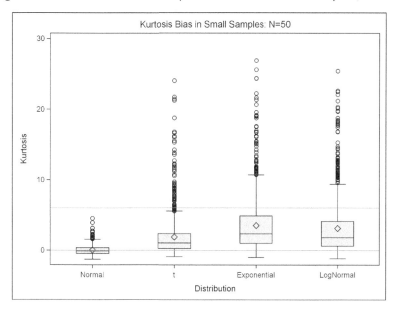

Figure 4.12 shows distributions of the sample kurtosis for small samples ($N = 50$) drawn from four different distributions. The mean value (the Monte Carlo estimate) is shown by a diamond. Horizontal reference lines are drawn to indicate the population kurtosis as shown in Table 4.1. For the normal distribution, the kurtosis sampling distribution is centered at 0, which is the value of the population kurtosis. However, the Monte Carlo estimates are less than the population value for the t, exponential, and lognormal distributions. Notice also that the distribution of the sample kurtosis is highly skewed and has a long tail.

The bias in the kurtosis estimate is less for large samples as shown in the following exercise.

Exercise 4.5: Repeat the simulation and redraw Figure 4.12 for samples that contain $N = 2000$ observations. Compare the range of sample kurtosis values for $N = 50$ and $N = 2000$.

Exercise 4.6: Is the skewness statistic also biased for small samples? Modify the example in this section to compute and plot the skewness of 1,000 random samples for $N = 50$. The skewness for each distribution is given in Table 4.1.

4.5   Simulating Data by Using the SAS/IML Language

This section shows how to simulate and analyze data by using the SAS/IML language. The first example simulates and analyzes univariate data; the second simulates multivariate data. The simulated samples and the ASD are analyzed by using SAS/IML functions.

4.5.1   The Sampling Distribution of the Mean Revisited

You can use SAS/IML software to repeat the simulation and computation in Section 4.4.3 and
Section 4.4.5. The following program computes an ASD for the sample mean of $U(0, 1)$ data
($N = 10$). Each sample is stored as a row of a matrix, **x**. This example shows an efficient way to
simulate and analyze many univariate samples in PROC IML. The results of the program are shown
in Figure 4.13.

```
%let N = 10;
%let NumSamples = 1000;
proc iml;
call randseed(123);
x = j(&NumSamples,&N);       /* many samples (rows), each of size N */
call randgen(x, "Uniform");  /* 1. Simulate data                    */
s = x[,:];                   /* 2. Compute statistic for each row   */
Mean = mean(s);              /* 3. Summarize and analyze ASD        */
StdDev = std(s);
call qntl(q, s, {0.05 0.95});
print Mean StdDev (q`)[colname={"5th Pctl" "95th Pctl"}];

/* compute proportion of statistics greater than 0.7 */
Prob = mean(s > 0.7);
print Prob[format=percent7.2];
```

Figure 4.13 Analysis of the ASD of the Sample Mean of $U(0, 1)$ Data, $N = 10$

Mean	StdDev	5th Pctl	95th Pctl
0.5026407	0.0925483	0.3540121	0.6588903

Prob
1.90%

Notice the following features of the SAS/IML program:

- There are no loops.

- Three statements are used to generate the samples: RANDSEED, J, and RANDGEN. A single
  call to the RANDGEN routine fills the entire matrix with random values.

In the program, the colon subscript reduction operator (:) is used to compute the mean of each row
of the **x** matrix. The column vector **s** contains the ASD. The mean, standard deviation, and quantile
of the ASD are computed by using the MEAN, STD, and QNTL functions, respectively. These
functions operate on each column of their matrix argument. (Other SAS/IML functions that operate
on columns include the MEDIAN and VAR functions.)

Figure 4.13 shows that the results are identical to the results in Figure 4.3 and Figure 4.6. The
sampling distribution (not shown) is identical to Figure 4.4 because both programs use the same seed
for the SAS random number generator and generate the same sequence of random variates. Notice
that the SAS/IML program is more compact than the corresponding analysis that uses the DATA step
and SAS procedures.

Notice that the PRINT statement prints the expression q`, which is read "q prime." The prime symbol is the matrix transpose operator. It converts a column vector into a row vector, or transposes an $N \times p$ matrix into a $p \times N$ matrix. You can also use the T function to transpose a matrix or vector.

Exercise 4.7: Rewrite the simulation so that each column of x is a sample. The column means form the ASD. Use the T function, which transposes a matrix, prior to computing the summary statistics.

Exercise 4.8: Use the SAS/IML language to estimate the sampling distribution for the maximum of 10 uniform random variates. Display the summary statistics.

4.5.2   Reshaping Matrices

Exercise 4.7 shows that you can store each simulated sample in the column of a matrix. This is sometimes called the "wide" storage format. This storage format is useful when you intend to analyze each sample in SAS/IML software.

An alternative approach is to generate the data in SAS/IML software but to write the data to a SAS data set that can be analyzed by some other procedure. In this approach, you need to reshape the simulated data into a long vector and manufacture a SampleID variable that identifies each sample. This is sometimes called the "long" storage format. To represent the data in the long format, use the REPEAT and SHAPE function to generate the **SampleID** variable (see Appendix A, "A SAS/IML Primer"), as shown in the following statements:

```
proc iml;
call randseed(123);
x = j(&NumSamples,&N);         /* many samples (rows), each of size N */
/* "long" format: first generate data IN ROWS... */
call randgen(x, "Uniform");        /* 1. Simulate data (all samples) */
ID = repeat( T(1:&NumSamples), 1, &N); /* {1   1 ...   1,
                                          2   2 ...   2,
                                        ... ... ... ...
                                        100 100 ... 100} */
/* ...then convert to long vectors and write to SAS data set */
SampleID = shape(ID, 0, 1);      /* 1 col, as many rows as necessary */
z = shape(x, 0, 1);
create Long var{SampleID z}; append; close Long;
```

The result of the SAS/IML program is a data set, Long, which is in the same format as the Sim data set that was created in Section 4.4.4. This trick works provided that the samples are stored in rows.

Actually, it is not necessary to reshape the matrices into vectors. The following statement also creates a data set with two columns. The data are identical to the Long data:

```
create Long2 var{ID x}; append; close Long2;
```

4.5.3   The Sampling Distribution of Pearson Correlations

When simulating multivariate data, each sample might be quite large. Rather than attempt to store all samples in memory, it is often useful to adopt a simulation scheme that requires less memory. In this scheme, only a single multivariate sample is ever held in memory.

To illustrate the multivariate approach to simulation in PROC IML, this example generates 20 random values from the bivariate normal distribution with correlation $\rho = 0.3$ by using the RANDNORMAL function, which is described in Section 8.3.1. Each call to the RANDNORMAL function returns a 20×2 matrix of ordered pairs from a correlated bivariate normal distribution. For each sample, the program computes the Pearson correlation between the variables.

```
%let N = 20;                        /* size of each sample */
%let NumSamples = 1000;             /* number of samples   */
proc iml;
call randseed(123);
mu = {0 0}; Sigma = {1 0.3, 0.3 1};
rho = j(&NumSamples, 1);            /* allocate vector for results   */
do i = 1 to &NumSamples;            /* simulation loop               */
    x = RandNormal(&N, mu, Sigma);  /* simulated data in N x 2 matrix */
    rho[i] = corr(x)[1,2];          /* Pearson correlation           */
end;
/* compute quantiles of ASD; print with labels */
call qntl(q, rho, {0.05 0.25 0.5 0.75 0.95});
print (q`)[colname={"P5" "P25" "Median" "P75" "P95"}];
```

Figure 4.14 Quantiles of the Approximate Sampling Distribution for a Pearson Correlation, $N = 20$

P5	P25	Median	P75	P95
-0.06808	0.1607904	0.3166604	0.455125	0.6424028

Each simulated sample consists of **N** observations and two columns and is held in the matrix **x**. Notice the DO loop. For each sample, the statistic (here, the correlation coefficient) is computed inside the DO loop and stored in the **rho** vector. During the next iteration, the **x** matrix is overwritten with a new simulated sample. This simulation scheme is often used for multivariate data. It requires minimal storage space (memory) but is less efficient because it is not vectorized as well as previous SAS/IML programs.

The result of this simulation shows that the sample correlation has a lot of variability for bivariate samples with 20 observations. As expected, the center of the sampling distribution is close to 0.3. However, 90% of the simulated correlations are in the interval $[-0.07, 0.64]$, which is quite wide.

Descriptive statistics are useful for summarizing the ASD of a statistic such as the correlation coefficient, but a histogram is often more revealing. The following statements write the **rho** variable to a data set and call PROC UNIVARIATE to plot the histogram:

```
create corr var {"Rho"}; append; close;      /* write ASD */
quit;

/* 3. Visualize approx. sampling distribution of statistic */
ods graphics on;
proc univariate data=Corr;
    label Rho = "Pearson Correlation Coefficient";
    histogram Rho / kernel;
    ods select Histogram;
run;
```

Figure 4.15 ASD of the Sample Correlation, Bivariate Data, $N = 20$, $\rho = 0.3$

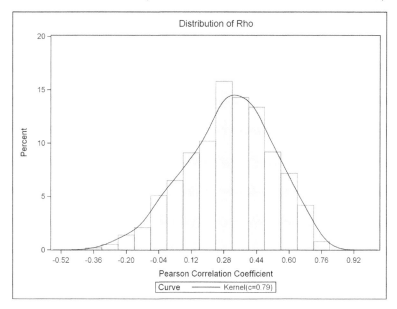

The distribution of the correlation coefficient is not symmetric for this example. It has negative skewness.

Exercise 4.9: Use the ASD to estimate the probability that the sample correlation coefficient is negative for a sample of size 20 from the bivariate normal distribution in this section. (See Section 4.4.3.)

4.6   References

Bai, J. and Ng, S. (2005), "Tests for Skewness, Kurtosis, and Normality for Time Series Data," *Journal of Business and Economic Statistics*, 23, 49–60.

Joanes, D. N. and Gill, C. A. (1998), "Comparing Measures of Sample Skewness and Kurtosis," *Journal of the Royal Statistical Society, Series D*, 47, 183–189.

Kendall, M. G. and Stuart, A. (1977), *The Advanced Theory of Statistics*, volume 1, 4th Edition, New York: Macmillan.

Ross, S. M. (2006), *Simulation*, 4th Edition, Orlando, FL: Academic Press.

Chapter 5
Using Simulation to Evaluate Statistical Techniques

5.1   Overview of Evaluating Statistical Techniques

This chapter describes how to use simulation to evaluate the performance of statistical techniques. The main idea is to use the techniques from Chapter 4, "Simulating Data to Estimate Sampling Distributions," to simulate many sets of data from a known distribution and to analyze the corresponding sampling distribution for a statistic. If you change the distribution of the population, then you also change the sampling distribution.

This chapter examines sampling distributions for several well-known statistics. For some statistics, the sampling distribution is only known asymptotically for large samples. This chapter applies simulation techniques to approximate the sampling distributions of these familiar statistics. The value of this chapter is that you can apply the same techniques to statistics for which the sampling distribution is unknown and to small samples where the asymptotic formulas do not apply. Often, simulation is the only technique for estimating the sampling distribution of an unfamiliar statistic.

This chapter describes how to use simulation to do the following:

- Estimate the coverage probability for the confidence interval for the mean of normal data. How does the coverage probability change when the data are nonnormal?

- Examine the performance of the standard two-sample t test. How does the test perform on data that do not satisfy the assumptions of the test?

- Estimate the power of the two-sample t test.

5.2   Confidence Interval for a Mean

A confidence interval is an interval estimate that contains the parameter with a certain probability. Each sample results in a different confidence interval. Due to sampling variation, the confidence interval that is constructed from a particular sample might not contain the parameter.

A 95% confidence interval is a statement about the probability that a confidence interval contains the parameter. In practical terms, a 95% confidence interval means that if you generate a large number of samples and construct the corresponding confidence intervals, then about 95% of the intervals will contain the parameter.

5.2.1   Coverage for Normal Data

You can use simulation to estimate the coverage probability of a confidence interval. Suppose you sample from the normal distribution. Let $\bar{x}$ be the sample mean and s be the sample standard deviation. The exact 95% confidence interval for the population mean is

$$[\bar{x} - t_{0.975,n-1}s/\sqrt{n}, \ \bar{x} + t_{0.975,n-1}s/\sqrt{n}]$$

where $t_{1-\alpha/2,n-1}$ is the $(1 - \alpha/2)$ quantile of the t distribution with $n - 1$ degrees of freedom (Ross 2006, p. 118). This computation is available in the MEANS procedure by using the LCLM= option and UCLM= option in the OUTPUT statement.

This section uses simulation of normal data to demonstrate that the coverage probability of the confidence interval is 0.95. A subsequent section violates the assumption of normality and explores what happens to the coverage probability if you use these intervals with nonnormal data.

The following DATA step generates 10,000 samples, each of size 50, from the standard normal distribution. The MEANS procedure computes the mean and 95% confidence interval for each sample. These sample statistics are stored in the OutStats data set.

```
%let N = 50;                             /* size of each sample   */
%let NumSamples = 10000;                 /* number of samples     */
/* 1. Simulate obs from N(0,1) */
data Normal(keep=SampleID x);
call streaminit(123);
do SampleID = 1 to &NumSamples;          /* simulation loop       */
   do i = 1 to &N;                       /* N obs in each sample  */
      x = rand("Normal");                /* x ~ N(0,1)            */
      output;
   end;
end;
run;
```

```
/* 2. Compute statistics for each sample */
proc means data=Normal noprint;
   by SampleID;
   var x;
   output out=OutStats mean=SampleMean lclm=Lower uclm=Upper;
run;
```

The SampleMean variable contains the mean for each sample; the Lower and Upper variables contain the left and right endpoints (respectively) of a 95% confidence interval for the population mean, which is 0 in this example. You can visualize these confidence intervals by stacking them in a graph. Figure 5.1 shows the first 100 intervals.

```
ods graphics / width=6.5in height=4in;
proc sgplot data=OutStats(obs=100);
   title "95% Confidence Intervals for the Mean";
   scatter x=SampleID y=SampleMean;
   highlow x=SampleID low=Lower high=Upper / legendlabel="95% CI";
   refline 0 / axis=y;
   yaxis display=(nolabel);
run;
```

Figure 5.1 Confidence Intervals for the Mean Parameter for 100 Samples from $N(0, 1)$, $N = 50$

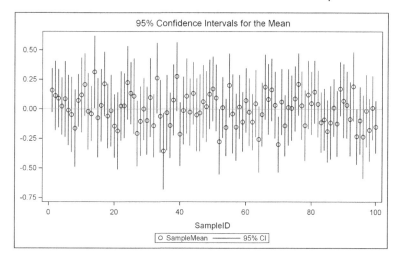

The graph shows that most of the confidence intervals contain zero, but a few confidence intervals do not contain zero. Those that do not contain zero correspond to "unlucky" samples in which an unusually large proportion of the data is negative or positive. By looking at the graph, you might be able to count how many of the 100 intervals do not contain zero, but for the 10,000 intervals, it is best to let the FREQ procedure do the counting. The following DATA step creates an indicator variable that has the value 1 for samples whose confidence interval contains zero, and has the value 0 otherwise:

```
/* how many CIs include parameter? */
data OutStats;  set OutStats;
   ParamInCI = (Lower<0 & Upper>0);               /* indicator variable */
run;
```

```
/* Nominal coverage probability is 95%. Estimate true coverage. */
proc freq data=OutStats;
   tables ParamInCI / nocum;
run;
```

Figure 5.2 Percentage of Confidence Intervals That Contain the Parameter

The FREQ Procedure

ParamInCI	Frequency	Percent
0	534	5.34
1	9466	94.66

For this simulation, 94.66% of the samples have a confidence interval that contains zero. For other values of the random seed, you might obtain 95.12% or 94.81%, but in general you should expect 95% of the confidence intervals to contain zero.

Exercise 5.1: Let P be the proportion of confidence intervals that contain zero. Rerun the program 10 times using 0 as the seed value, and record the range of values of P that you observe. Reduce the number of samples to 1,000 and run the new program 10 times. Compare the range of P values. Explain what you observe.

5.2.2   Coverage for Nonnormal Data

If the data are not normally distributed, then the interval $[\bar{x} - t_{0.975,n-1}s/\sqrt{n}, \bar{x} + t_{0.975,n-1}s/\sqrt{n}]$ contains the population mean with a probability that is different from 0.95. To demonstrate this, consider sampling from the standard exponential distribution, which has mean 1. In order to reuse the code from the previous section, subtract 1 from each observation in the sample so that the expected value of the population is zero, as shown here:

```
data Exp;
   ...
   x = rand("Expo") - 1;              /* x ~ Exp(1) - 1. Note E(X)=0 */
   ...
run;
```

You can now use the Exp data set as the input for the MEANS procedure in the previous program. The percentage of confidence intervals that contain zero (the true mean) is no longer close to 95% (relative to the standard error) as shown in Figure 5.3.

Figure 5.3 Number of Intervals That Contain the True Mean for Exponential Data

The FREQ Procedure

ParamInCI	Frequency	Percent
0	656	6.56
1	9344	93.44

For data drawn from the exponential data, the coverage probability is less than 95%. Simulation makes it easy to see that different data distributions affect the results of statistical methods.

Statisticians have developed exact confidence intervals for nonnormal distributions, including the exponential (Hahn and Meeker 1991).

Exercise 5.2: Produce Figure 5.3 by completing the details in this section.

Exercise 5.3: Use the BINOMIAL option on the TABLES statement to show that a 95% confidence interval about the estimate of 0.9344 does not include 0.95.

5.2.3 Computing Coverage in the SAS/IML Language

Section 5.2.1 uses simulation to estimate the coverage probability of the exact 95% confidence interval for normally distributed data. The following SAS/IML program repeats the analysis as a way of showing how to use the SAS/IML language for statistical simulation:

```
%let N = 50;                         /* size of each sample */
%let NumSamples = 10000;             /* number of samples   */
proc iml;
call randseed(321);
x = j(&N, &NumSamples);              /* each column is a sample   */
call randgen(x, "Normal");          /* x ~ N(0,1)                */

SampleMean = mean(x);               /* mean of each column       */
s = std(x);                         /* std dev of each column    */
talpha = quantile("t", 0.975, &N-1);
Lower = SampleMean - talpha * s / sqrt(&N);
Upper = SampleMean + talpha * s / sqrt(&N);

ParamInCI = (Lower<0 & Upper>0);    /* indicator variable        */
PctInCI = ParamInCI[:];             /* pct that contain parameter */
print PctInCI;
```

Most statements are explained in the program comments. The RANDGEN routine generates a $50 \times 10,000$ matrix of standard normal variates. Each column is a sample. The MEAN and STD functions compute the mean and standard deviation, respectively, of each column. The QUANTILE function, which is described in Section 3.2, is used to compute the $1 - 0.05/2$ quantile of the t distribution. Notice that **SampleMean**, **s**, **Lower**, and **Upper** are $1 \times 10,000$ row vectors. Consequently, **ParamInCI** is a $1 \times 10,000$ vector of zeros and ones. The colon operator (:) is used to compute the mean of the indicator variable, which is the proportion of ones. The results are displayed in Figure 5.4.

Figure 5.4 Proportion of Confidence Intervals That Contain the Parameter

PctInCI
0.9488

Of the 10,000 samples of size 10, about 95% of them produce confidence intervals that contain zero.

Exercise 5.4: Repeat the SAS/IML analysis for exponentially distributed data.

5.3   Assessing the Two-Sample t Test for Equality of Means

A simple statistical test is the two-sample t test for the population means of two groups. There are several variations of the t test, but the *pooled variance* t test that is used in this section has three assumptions:

- The samples are drawn independently from their respective populations.

- The populations are normally distributed.

- The variances of the populations are equal.

In short, the t test assumes that samples are drawn independently from $N(\mu_1, \sigma)$ and $N(\mu_2, \sigma)$. The pooled variance t test enables you to use the sample data to test the null hypothesis that $\mu_1 = \mu_2$.

This section examines the robustness of the two-sample pooled-variance t test to the second and third assumptions. Similar analyses were conducted by Albert (2009, p. 12), Bailer (2010, p. 290), and Fan et al. (2002, p. 118).

5.3.1   Robustness of the t Test to Unequal Variances

This section uses simulation to study the sensitivity of the pooled-variance t test to the assumption of equal variances. The study consists of simulating data that satisfy the null hypothesis ($\mu_1 = \mu_2$) for the test. Consequently, it is a Type I error if the t test rejects the null hypothesis for these data. (Recall that a Type I error occurs when a null hypothesis is falsely rejected.)

For the remainder of this study, fix the significance level of the test at $\alpha = 0.05$. Consider two scenarios. In the first scenario, the data are simulated from two $N(0, 1)$ populations. For these data, the populations have equal variances, and the probability of a Type I error is α. In the second scenario, one sample is drawn from an $N(0, 1)$ distribution whereas the other is drawn from $N(0, 10)$. Because the populations have different variances, the actual Type I error rate should be different from α.

The following DATA step generates 10,000 samples (each of size 10) from two populations. The classification variable, c, identifies the populations. The x1 variable contains the data for the first scenario; the x2 variable contains the data for the second scenario.

```
/* test sensitivity of t test to equal variances */
%let n1 = 10;
%let n2 = 10;
%let NumSamples = 10000;                    /* number of samples       */

/* Scenario 1: (x1 | c=1) ~ N(0,1);   (x1 | c=2) ~ N(0,1);            */
/* Scenario 2: (x2 | c=1) ~ N(0,1);   (x2 | c=2) ~ N(0,10);           */
data EV(drop=i);
label x1 = "Normal data, same variance"
      x2 = "Normal data, different variance";
call streaminit(321);
do SampleID = 1 to &NumSamples;
   c = 1;                                   /* sample from first group  */
   do i = 1 to &n1;
      x1 = rand("Normal");
      x2 = x1;
      output;
   end;
   c = 2;                                   /* sample from second group */
   do i = 1 to &n2;
      x1 = rand("Normal");
      x2 = rand("Normal", 0, 10);
      output;
   end;
end;
run;
```

Notice the structure of the simulated data. The first 10 observations are for Group 1 ($c = 1$), the next 10 observations are for Group 2 ($c = 2$), and this pattern repeats as shown in Table 5.1. This structure is chosen because it also enables you to simulate data for which the number of observations in the two groups are not equal (see Exercise 5.7).

Table 5.1   Structure of the Simulated Data

Observation	SampleID	c	x1	x2
1	1	1	$x1_1$	$x2_1$
$\vdots$	$\vdots$	$\vdots$	$\vdots$	$\vdots$
10	1	1	$x1_{10}$	$x2_{10}$
11	1	2	$x1_{11}$	$x2_{11}$
$\vdots$	$\vdots$	$\vdots$	$\vdots$	$\vdots$
20	1	2	$x1_{20}$	$x2_{20}$
21	2	1	$x1_{21}$	$x2_{21}$
$\vdots$	$\vdots$	$\vdots$	$\vdots$	$\vdots$
30	2	1	$x1_{30}$	$x2_{30}$
31	2	2	$x1_{31}$	$x2_{31}$
$\vdots$	$\vdots$	$\vdots$	$\vdots$	$\vdots$
40	2	2	$x1_{40}$	$x2_{40}$
$\vdots$	$\vdots$	$\vdots$	$\vdots$	$\vdots$

In order to conduct the standard two-sample t test, you can call the TTEST procedure and use a BY statement to analyze each pair of samples. For each BY group, the TTEST procedure produces default output. However, to prevent thousands of pages of output, you should use ODS to suppress the output during the computation of the t statistics. In the following statements, the %ODSOFF macro is used to suppress output. The macro is described in Section 6.4.2. The ODS OUTPUT statement is used to write a data set that contains the p-value for each test.

```
/* 2. Compute statistics */
%ODSOff                             /* suppress output              */
proc ttest data=EV;
   by SampleID;
   class c;                         /* compare c=1 to c=2           */
   var x1-x2;                       /* run t test on x1 and also on x2 */
   ods output ttests=TTests(where=(method="Pooled"));
run;
%ODSOn                              /* enable output                */
```

As shown in Section 5.2, you can use a DATA step to create a binary indicator variable that has the value 1 if the t test on the sample rejects the null hypothesis, and the value 0 if the t test does not reject the null hypothesis. You can use PROC FREQ to summarize the results, which are shown in Figure 5.5.

```
/* 3. Construct indicator var for tests that reject H0 at 0.05 significance */
data Results;
   set TTests;
   RejectH0 = (Probt <= 0.05);          /* H0: mu1 = mu2           */
run;

/* 3b. Compute proportion: (# that reject H0)/NumSamples */
proc sort data=Results;
   by Variable;
run;

proc freq data=Results;
   by Variable;
   tables RejectH0 / nocum;
run;
```

Figure 5.5 Summary of t Tests on 10,000 Samples, Normal Data

The FREQ Procedure

Variable=x1

RejectH0	Frequency	Percent
0	9457	94.57
1	543	5.43

Figure 5.5 *continued*

The FREQ Procedure

Variable=x2

RejectH0	Frequency	Percent
0	9355	93.55
1	645	6.45

Figure 5.5 summarizes the results of the two scenarios. For the x1 variable (equal population variances), the *t* test rejects the null hypothesis for about 5% of the samples as predicted by theory. You can use the BINOMIAL option in the TABLES statement in PROC FREQ to assess the uncertainty in the simulation study. (See Exercise 5.5.)

For the x2 variable (unequal variances), 6.5% of the samples reject the null hypothesis. You should do additional simulation before you draw a conclusion, but this one simulation indicates that the *t* test rejects the null hypothesis more than 5% of the time when the population variances are not equal. Consequently, it is wise to use the Satterthwaite test rather than the pooled-variance test if you suspect unequal variances.

Exercise 5.5: Use the BINOMIAL option in the TABLES statement to compute a confidence interval for the proportion that the *t* test rejects the null hypothesis.

Exercise 5.6: The relative magnitudes of the population variances determine the proportion of samples in which the *t* test rejects the null hypothesis. Rerun the simulation when x2 (for c=2) is drawn from the $N(0, 2)$, $N(0, 5)$, and $N(0, 100)$ distributions. How sensitive is the pooled-variance *t* test to differences in the population variances?

Exercise 5.7: Rerun the simulation when the first sample (c=1) contains 20 observations and the second sample contains 10 observations. Compare the results with the results of a simulation in which both samples contain 15 observations. Which test has more power?

5.3.2   Robustness of the *t* Test to Nonnormal Populations

How sensitive is the two-sample pooled-variance *t* test to the assumption that both populations are normally distributed?

You can modify the program from Section 5.3.1 to sample from nonnormal distributions. For example, in one scenario you can draw samples from an exponential distribution for both the first group (c=1) and the second group (c=2). In another scenario, you can choose the first group from the normal distribution with unit standard deviation and the second group from an exponential distribution with a standard deviation 10 times as large. In each scenario, choose the population distributions so that the null hypothesis is true.

```
   . . .
   c = 1;
   do i = 1 to &n1;
       x3 = rand("Exponential");            /* mean = StdDev = 1   */
       x4 = rand("Normal", 10);             /* mean=10; StdDev = 1 */
       output;
   end;
   c = 2;
   do i = 1 to &n2;
       x3 = rand("Exponential");            /* mean = StdDev = 1   */
       x4 = 10 * rand("Exponential");       /* mean = StdDev = 10  */
       output;
   end;
   . . .
```

Figure 5.6 shows the results of *t* tests on the 10,000 samples. For the first scenario (both populations are Exp(1)), the test rejects the null hypothesis about 4.3% of the time. For the second scenario (one group is normal, the other is exponential, and the variances are unequal), the test rejects the null hypothesis about 11.4% of the time.

Figure 5.6 Summary of *t* Tests on 10,000 Samples, Nonnormal Data

The FREQ Procedure

Variable=x3

RejectH0	Frequency	Percent
0	9573	95.73
1	427	4.27

The FREQ Procedure

Variable=x4

RejectH0	Frequency	Percent
0	8857	88.57
1	1143	11.43

One of the drawbacks of simulation is that it is difficult to generalize the results. Although it is clear that the *t* test rejects the null hypothesis for a large number of samples for the x4 variable, the simulation does not indicate *why* this is so. Is there something special about the exponential distribution, or would using a gamma or lognormal distribution give similar results? For the other tests, the distributions for each group are from the same family, but for x4 one sample is normally distributed whereas the other is exponentially distributed. Does that matter? Chapter 6, "Strategies for Efficient and Effective Simulation," discusses these issues.

Exercise 5.8: Repeat the simulation for a new variable x5, which is drawn from a gamma(10) distribution when $c = 1$ and an exponential(10) distribution for $c = 2$. Both of these distributions have the same mean and variance, but both distributions are nonnormal. What percentage of the time does the *t* test reject the null hypothesis?

5.3.3   Assessing the *t* Test in SAS/IML Software

You can also investigate the robustness of the *t* test to its assumptions by using SAS/IML software. The formulas for the two-sample pooled variance *t* test are given in the documentation for the TTEST procedure in the *SAS/STAT User's Guide*. The following SAS/IML program reproduces the simulation of the x4 variable from Section 5.3.2 in which the first population is a normal distribution with unit standard deviation, and the second population is an exponential distribution with a standard deviation 10 times as large:

```
%let n1 = 10;
%let n2 = 10;
%let NumSamples = 1e4;                    /* number of samples */

proc iml;
/* 1. Simulate the data by using RANDSEED and RANDGEN, */
call randseed(321);
x = j(&n1, &NumSamples);                  /* allocate space for Group 1 */
y = j(&n2, &NumSamples);                  /* allocate space for Group 2 */
call randgen(x, "Normal", 10);            /* fill matrix from N(0,10)   */
call randgen(y, "Exponential");           /* fill from Exp(1)           */
y = 10 * y;                               /* scale to Exp(10)           */

/* 2. Compute the t statistics; VAR operates on columns */
meanX = mean(x);   varX = var(x);         /* mean & var of each sample  */
meanY = mean(y);   varY = var(y);
/* compute pooled standard deviation from n1 and n2 */
poolStd = sqrt( ((&n1-1)*varX + (&n2-1)*varY)/(&n1+&n2-2) );

/* compute the t statistic */
t = (meanX - meanY) / (poolStd*sqrt(1/&n1 + 1/&n2));

/* 3. Construct indicator var for tests that reject H0 */
alpha = 0.05;
RejectH0 = (abs(t)>quantile("t", 1-alpha/2, &n1+&n2-2));   /* 0 or 1 */

/* 4. Compute proportion: (# that reject H0)/NumSamples */
Prob = RejectH0[:];
print Prob;
```

Figure 5.7 SAS/IML Summary of *t* Tests on 10,000 Samples, Nonnormal Data

Prob
0.1091

In this program, the formula for the *t* test is explicitly used so that you can see how the SAS/IML language supports a natural syntax for implementing statistical methods. In the first part of the program, the RANDGEN subroutine generates all samples in a single call. Each column of the **x** matrix is a normal sample with 10 observations (rows). Similarly, each column of the **y** matrix is exponentially distributed. Notice that the SAS/IML approach does not use a classification variable, **c**. Instead, the observations with **c=1** are stored in the matrix **x**, and the observations with **c=2** are stored in the matrix **y**.

The second part of the program computes the sample statistics and stores them in the $1 \times 10{,}000$ vector $\mathbf{t}$. All of the variables in this section are vectors. If you plan to use t tests often in PROC IML, then you can encapsulate these statements into a user-defined function.

The third part of the program compares the sample statistics with the $(1 - \alpha/2)$ quantile of the t distribution in order to generate an indicator variable, `RejectH0`. The colon subscript reduction operator (:) is used to find the mean of this vector, which is the proportion of elements that contains 1. This estimates the Type I error rate for the t test given the distribution of the populations. The error rate, which is displayed in Figure 5.7, is approximately 11%.

5.3.4 Summary of the t Test Simulations

The previous sections use simulation to examine the robustness of the two-sample pooled t test to normality and to equality of variances. The results indicate that the two-sample pooled t test is relatively robust to departures from normality but is sensitive to (large) unequal variances.

You can use the ideas in this chapter (and in the exercises) to build a large, systematic, simulation study that more carefully examines how the t test behaves when its assumptions are violated. Chapter 6 provides strategies for designing large simulation studies.

5.4   Evaluating the Power of the t Test

The simulations so far have explored how the two-sample t test performs when both samples are from populations with a common mean. That is, the simulations sample from populations that satisfy the null hypothesis of the test, which is $\mu_1 = \mu_2$, where μ_1 and μ_2 are the population means. A related question is, "How good is the t test at detecting differences in the population means?"

Suppose that $\mu_2 = \mu_1 + \delta$, and, for simplicity, assume that the two populations are normally distributed with equal variance. When $\delta = 0$, the population means are equal and the standard t test (with significance level $\alpha = 0.05$) rejects the null hypothesis 5% of the time. If δ is larger than zero, then the t test should reject the null hypothesis more often. As δ gets quite large (relative to the standard deviation of the populations), the t test should reject the null hypothesis for almost every sample.

5.4.1   Exact Power Analysis

The *power* of the t test is the probability that the test will reject a null hypothesis that is, in fact, false. The power of the two-sample t test can be derived theoretically. In fact, the POWER procedure in SAS/STAT software can compute a "power curve" that shows how the power varies as a function of δ for normally distributed data:

```
proc power;
   twosamplemeans  power = .             /* missing ==> "compute this" */
      meandiff= 0 to 2 by 0.1            /* delta = 0, 0.1, ..., 2      */
      stddev=1                           /* N(delta, 1)                 */
      ntotal=20;                         /* 20 obs in the two samples   */
   plot x=effect markers=none;
   ods output Output=Power;              /* output results to data set */
run;
```

Figure 5.8 shows that when the means of the two groups of size 10 are two units apart (twice the standard deviation), the *t* test is almost certain to detect the difference in means. For means that are one unit apart (equal to the standard deviation), the chance of detecting the difference is about 56%. Of course, if the samples contain more than 10 observations, then the test can detect much smaller differences.

Figure 5.8 Power Curve for Two-Sample *t* Test, $N_1 = N_2 = 10$

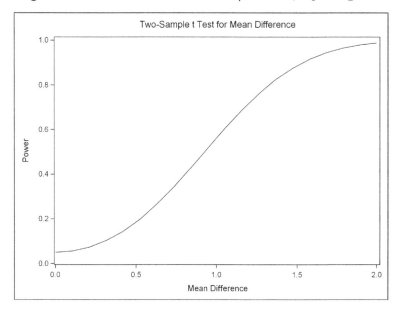

5.4.2   A Simulated Power Analysis

Theoretical results for power computations are not always available. However, you can use simulation to approximate a power curve. This section uses simulation to estimate points along the power curve in Figure 5.8. To do this, use the techniques from Section 5.3 to sample from $N(0, 1)$ and $N(\delta, 1)$ as δ varies from 0 to 2, and estimate the power for each value of δ. In particular, add a loop that iterates over the values of δ, as follows:

```
do Delta = 0 to 2 by 0.1;
   do SampleID = 1 to &NumSamples;
      ...
      x1 = rand("Normal");                /* for c=1 */
      ...
      x1 = rand("Normal", Delta, 1);   /* for c=2 */
      ...
   end;
end;
```

Modify the call to PROC TTEST by adding the **Delta** variable to the list of BY variables:

```
by Delta SampleID;
```

You can then use PROC FREQ to estimate the power for each value of **Delta**. You can also combine these simulated estimates with the exact power values that are produced by PROC POWER in Section 5.4.1:

```
proc freq data=Results noprint;
   by Delta;
   tables RejectH0 / out=SimPower(where=(RejectH0=1));
run;

/* merge simulation estimates and values from PROC POWER */
data Combine;
   set SimPower Power;
   p = percent / 100;
   label p="Power";
run;

proc sgplot data=Combine noautolegend;
   title "Power of the t Test";
   title2 "Samples are N(0,1) and N(delta,1), n1=n2=10";
   series x=MeanDiff y=Power;
   scatter x=Delta y=p;
   xaxis label="Difference in Population Means (mu2 - mu1)";
run;
```

Figure 5.9 shows that the estimates obtained by the simulation with 10,000 samples are close to the exact values that are produced by PROC POWER. The simulation that creates Figure 5.9 is computationally intensive. It requires three nested loops in the DATA step and produces $21 \times 10{,}000 \times 20 = 4{,}200{,}000$ observations and 210,000 t tests. Nevertheless, the entire simulation requires less than 30 seconds on a desktop PC that was manufactured in 2010.

Figure 5.9 Power Curve and Estimates from Simulation

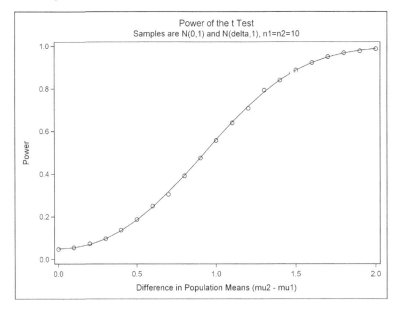

Exercise 5.9: In Figure 5.9, each estimate is the proportion of times that a binary variable, `RejectH0`, equals 1. Use the BINOMIAL option in the TABLES statement in PROC FREQ to compute 95% confidence intervals for the proportion. Add these confidence intervals to Figure 5.9 by using the YERRORLOWER= option and YERRORUPPER= option in the SCATTER statement.

5.5   Effect of Sample Size on the Power of the *t* Test

The simulation in the previous section examines the power of the *t* test to reject the null hypothesis $\mu_1 = \mu_2$ for various magnitudes of the difference $|\mu_1 - \mu_2|$. Throughout the simulation, the sample sizes n_1 and n_2 are held constant. However, you can examine the power for various choices of the sample size for a constant magnitude of the difference between means, which is known as the *effect size*.

To be specific, suppose that the first group is a control group, whereas the second group is an experimental group. The researcher thinks that a treatment can increase the mean of the second group by 0.5. The researcher wants to investigate how large the sample size should be in order to have a power of 0.8. In other words, assuming that $\mu_2 = \mu_1 + 0.5$, what should the sample size be so that the null hypothesis is rejected (at the 5% significance level) 80% of the time?

This sort of simulation occurs often in clinical trials. Computationally, it is very similar to the previous power computation, and once again PROC POWER can provide a theoretical power curve. However, for many situations that are encountered in practice, theoretical results are not available and simulation provides a way to obtain the power curve as a function of the sample size.

For convenience, assume that both groups have the same sample size, $n_1 = n_2 = N$. The following simulation samples N observations from $N(0, 1)$ and $N(0.5, 1)$ populations for $N = 40, 45, \ldots, 100$. For each value of the sample size, 1,000 samples are generated.

```
/* The null hypothesis for the t test is H0: mu1 = mu2.
   Assume that mu2 = mu1 + delta.
   Find sample size N that rejects H0 80% of the time.   */
%let NumSamples = 1000;               /* number of samples */

data PowerSizeSim(drop=i Delta);
call streaminit(321);
Delta = 0.5;                          /* true difference between means */
do N =   40 to 100 by 5;              /* sample size                   */
   do SampleID = 1 to &NumSamples;
      do i = 1 to N;
         c = 1; x1 = rand("Normal");              output;
         c = 2; x1 = rand("Normal", Delta, 1); output;
      end;
   end;
end;
run;
```

This is a coarse grid of values for N. It is usually a good idea to do a small scale simulation on a coarse grid (and with a moderate number of samples) in order to narrow down the range of sample sizes. The remainder of the program is the same as in Section 5.4.2, except that you use the N variable in the BY statements instead of the Delta variable. The results are shown in Figure 5.10.

Figure 5.10 Power versus Sample Size

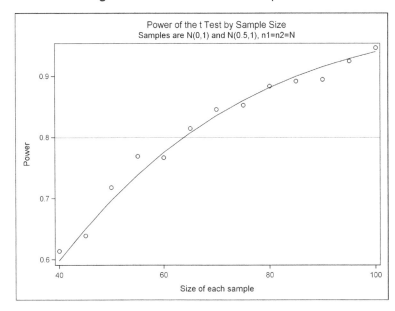

The markers in Figure 5.10 are the estimates of power obtained by the simulation. (You could also add binomial confidence intervals to the plot.) For this simple example, the exact results can be computed by using PROC POWER, so the exact power curve is overlaid on the estimates.

From an initial simulation on a coarse grid, about 65 subjects in each group are needed to obtain 80% power to detect a mean difference of 0.5. After you have narrowed down the possible range of sample sizes, you can decide whether you need to run a second, more accurate, simulation on a finer grid. For example, you might run a second simulation with $N = 60, 61, \ldots, 68$ and with 10,000 samples for each value of N.

Exercise 5.10: Run a second simulation on a finer grid, using 10,000 samples for each value of N.

5.6   Using Simulation to Compute *p*-Values

An important use of simulations is the computation of *p*-values in hypothesis testing. To illustrate this technique, consider an experiment in which a six-sided die is tossed 36 times. The number of times that each face appeared is shown in Table 5.2:

Table 5.2 Frequency Count for Each Face of a Tossed Die

Face	1	2	3	4	5	6
Count	8	4	4	3	6	11

Suppose that you want to determine whether this die is fair based on these observed frequencies. The chi-square test is the usual way to test whether a set of observed frequencies are consistent with a specified theoretical distribution. For a given set of observed counts, N_i, the chi-square statistic is the quantity $Q = \sum_{i=1}^{6} (N_i - Np_i)^2 / Np_i$, where $N = \sum_{i=1}^{6} N_i = 36$.

To test whether the die is fair, the null hypothesis for the chi-square test is that the probability of each face appearing is $1/6$. That is, $P(X = i) = p_i = 1/6$ for $i = 1, 2, \ldots, 6$. Consequently, the observed value of the statistic for the data in Table 5.2 is $q = 7.667$.

The p-value is the probability under the null hypothesis that the random variable Q is greater than or equal to the observed value: $P_{H_0}(Q \geq q)$. For large samples, the Q statistic has a χ^2 distribution with $k - 1$ degrees of freedom, where k is the number of categories. (For a six-sided die, $k = 6$.) For small samples, you can use simulation to approximate the sampling distribution of Q under the null hypothesis. You can then estimate the p-value as the proportion of times that the test statistic, which is computed for simulated data, is greater than the observed value.

You can perform this computation using the SAS DATA step and the FREQ procedure, which performs chi-square tests. However, you can also perform the simulation by writing a SAS/IML program. The SAS/IML language supports the RANDMULTINOMIAL function, which generates random samples from a multinomial distribution (see Section 8.2). The counts for each face of a tossed die can be generated as a draw from the multinomial distribution: Each draw is a set of six integers, where the ith integer represents the number of times that the ith face appeared out of 36 tosses.

The SAS/IML implementation of the simulation is very compact:

```
proc iml;
Observed = {8 4 4 3 6 11};                     /* observed counts */
k = ncol(Observed);                            /*  6              */
N = sum(Observed);                             /*  36             */
p = j(1, k, 1/k);                              /* {1/6,...,1/6}   */
Expected = N*p;                                /* {6,6,...,6}     */
qObs = sum( (Observed-Expected)##2/Expected ); /* q               */

/* simulate from null hypothesis */
NumSamples = 10000;
counts = RandMultinomial(NumSamples, N, p);    /* 10,000 samples  */
Q = ((counts-Expected)##2/Expected )[ ,+];     /* sum each row    */
pval = sum(Q>=qObs) / NumSamples;              /* proportion > q  */
print qObs pval;
```

Figure 5.11 Observed Value of χ^2 Statistic and p-Value

qObs	pval
7.6666667	0.1797

Figure 5.11 shows the observed value of the statistic and the proportion of simulated statistics (which were generated under the null hypothesis) that exceed the observed value. The simulation indicates that about 17% of the simulated samples result in a test statistic that exceeds the observed value. Consequently, the observed statistic is not highly unusual, and there is not sufficient evidence to reject the hypothesis of a fair die.

You can visualize the distribution of the test statistic and see where the observed statistic is located. The previous SAS/IML program did not end with a QUIT statement, so the following statements continue the program:

```
call symputx("qObs", qObs);                /* create macro variables */
call symputx("pval", pval);
create chi2 var {Q}; append; close chi2;
quit;
```

The statements use the SYMPUTX subroutine to create two macro variables that contain the value of **qObs** and **pval**, respectively. The simulated statistics are then written to a SAS data set. You can use the SGPLOT procedure to construct a histogram of the distribution, and you can use the REFLINE statement to indicate the location of the observed statistic. See Figure 5.12.

```
proc sgplot data=chi2;
    title "Distribution of Test Statistic under Null Hypothesis";
    histogram Q / binstart=0 binwidth=1;
    refline &qObs / axis=x;
    inset "p-value = &pval";
    xaxis label="Test Statistic";
run;
```

Figure 5.12 Approximate Sampling Distribution of the Test Statistic and Observed Value

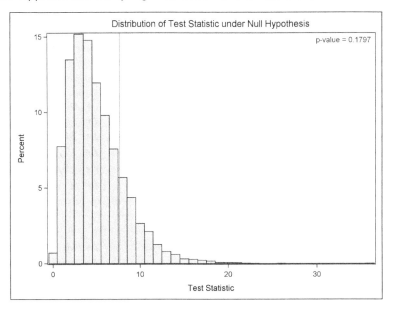

If an observed statistic is in the extreme tail of the distribution, then the *p*-value is small. The INSET statement was used to include the value of the *p*-value in Figure 5.12.

Exercise 5.11: Construct a SAS data set with the data in Table 5.2 and run PROC FREQ. The CHISQ option in the TABLES statement conducts the usual chi-square test. Compare the *p*-value from the simulation to the *p*-value computed by PROC FREQ.

Exercise 5.12: Simulate 10,000 samples of 36 random values drawn uniformly from $\{1, 2, 3, 4, 5, 6\}$. Use PROC FREQ with BY-group processing to compute the chi-square statistic for each sample. Output the OneWayChiSq table to a SAS data set. Construct a histogram of the chi-square statistics, which should look similar to Figure 5.12.

5.7 References

Albert, J. H. (2009), *Bayesian Computation with R*, 2nd Edition, New York: Springer-Verlag.

Bailer, J. (2010), *Statistical Programming in SAS*, Cary, NC: SAS Institute Inc.

Fan, X., Felsovályi, A., Sivo, S. A., and Keenan, S. C. (2002), *SAS for Monte Carlo Studies: A Guide for Quantitative Researchers*, Cary, NC: SAS Institute Inc.

Hahn, G. J. and Meeker, W. Q. (1991), *Statistical Intervals: A Guide for Practitioners*, New York: John Wiley & Sons.

Ross, S. M. (2006), *Simulation*, 4th Edition, Orlando, FL: Academic Press.

Chapter 6
Strategies for Efficient and Effective Simulation

Contents

6.1 Overview of Simulation Strategies

An *efficient* simulation is one that optimizes computational resources so that it finishes in a reasonable amount of time. An *effective* simulation is one that produces a desired statistical result. Efficiency is achieved through good programming practices and is a primary theme of this book. Effectiveness is achieved though a solid understanding of statistical concepts and methodology.

The design of a simulation study affects its efficiency and effectiveness. One design choice is the number of samples to use in a simulation. If you use too few samples, then the program runs quickly but uncertainty in the estimates renders the simulation useless. If you use too many samples, then the standard error of the estimates are tiny but the program requires hours or days to finish. Clearly, the best choice lies between these two extremes.

This chapter discusses issues that are important for designing efficient and effective simulation studies. The ideas in this chapter are applied throughout this book, but this chapter collects the various strategies in one place. This chapter also describes details of implementing simulations in SAS software, and points out why some implementations are more efficient than others.

6.2   The Design of a Simulation Study

Gentle (2009, Appendix A) discusses the fact that simulation studies are computer experiments. As such, Gentle argues, the principles of statistical design can be used to improve the efficiency, reproducibility, and documentation of a simulation study.

Gentle outlines a typical simulation study. Suppose that you want to study the power of ordinary least squares regression to test the hypothesis $H_0: \beta_1 = 0$ for the simple regression model $Y_i = \beta_0 + \beta_1 X_i + E_i$, where each response value, Y_i, is modeled as a linear function of an explanatory variable value, X_i, and a random variable, E_i, which represents an error term.

This problem contains many experimental factors that might or might not be important in computing the power. Among these factors are the following:

- The true value of β_1. You might choose equally spaced values, such as 0, 0.2, 0.4, ...

- The sample size. You might choose three sample sizes, such as 20, 200, and 2,000.

- The distribution of the error term. You might choose five distributions: a standard normal distribution and four contaminated normal distributions (see Section 7.5.2) with various percentages and magnitudes of contamination.

- The design points of the independent variable, x. You might choose three designs: points that are equally spaced, points that have a cluster of extreme values, and points drawn randomly from a skewed distribution.

If you use a full factorial design with, for example, 16 values for β_1, then there are $16 \times 3 \times 5 \times 3 = 720$ experiments that you need to run. Each experiment is a simulation study, which consists of generating thousands of random samples and running a regression analysis on each sample. It is tempting to feel overwhelmed by the magnitude of the study and by the challenge of summarizing the results. How should you proceed?

The best advice is to start small and start simply. For this experiment, you might want to choose the smallest sample size (20) and the simplest choice for the design of x (equally spaced) and for the distribution of the error term (normally distributed). You should also use a smaller number of samples than the number that you intend to use for the final study. One-tenth the number is often appropriate during the development, debugging, profiling, and optimization of the simulation algorithm.

When you run the small-scale simulation, consider whether you can present the results graphically. For example, Figure 6.1 presents the results as a curve that shows the power as a function of the parameter β_1. Standard errors for the estimates are displayed. Section 11.4 shows how to create graphs like Figure 6.1.

Figure 6.1 Preliminary Results of a Simulation Study

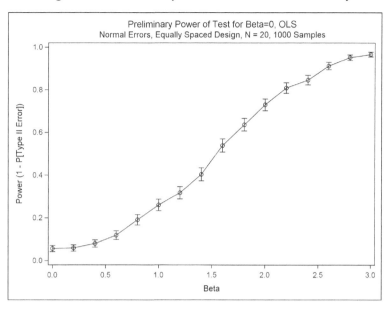

You can learn a lot from small-scale simulations, such as the values of β_1 for which the power curve is steep or flat, and the magnitude of the standard error for the estimates along the curve. Use this preliminary analysis to plan the final simulation. For example, use the preliminary analysis to choose an appropriate range for β_1. For small samples, you need a large range for β_1; for large samples, you need a much smaller range. Because the preliminary analysis estimates the standard error along the curve, you might decide to generate more samples when the variance of the estimator is relatively large (approximately $1 \leq \beta_1 \leq 2$ in Figure 6.1) and use fewer samples when the variance is smaller.

You can also use the preliminary analysis to estimate the run time for the final simulation. If you used 1,000 samples for each simulation in the preliminary study, then a simulation that uses 10,000 samples will take about 10 times as long.

In a similar way, you can run preliminary analyses for some of the other parameter values and scenarios, such as for a contaminated normal error term and for different design points for x. For each scenario, you should save the results (the points and upper/lower limits of the power curve) so that, for example, you can combine several curves on a single graph. By the time you run your final simulation study, you should have a good idea what the results will be.

In summary, when you begin a large simulation study, resist the urge to immediately write and run the entire simulation. Start small. Debug and optimize the simple cases. Run a small-scale simulation. These techniques provide important information that you can incorporate into the final study.

6.3   The Effect of the Number of Samples

There are two "sizes" in a simulation study, and it is important to not confuse them. One is the sample size; the other is the number of samples, which is also known as the number of *repetitions*.

As demonstrated in Section 4.4.5, the sample size, N, affects the width of the sampling distribution (that is, the standard error of the statistic). For most well-behaved statistics that are used in practice, small sample sizes correspond to large standard errors, whereas large sample sizes result in smaller standard errors. The value of N is often controlled in a simulation study. It might be set to the sample size of an observed data set, or it might be systematically varied, such as in the power study that is described in Section 6.2.

In contrast, the number of samples in the simulation is something that you have to choose every time that you run a simulation. The number of samples (which in this book is determined by the `NumSamples` macro variable) determines how well the approximate sampling distribution (ASD) of a statistic approximates the exact sampling distribution.

How many samples are sufficient? Unfortunately, that is a difficult question to answer. The number of samples that you need depends on characteristics of the sampling distribution. Lower order moments of the sampling distribution (such as the mean) require fewer samples than statistics that are functions of higher order moments, such as the variance and skewness. You might need many, many, samples to capture the extreme tail behavior of a sampling distribution.

A popular choice in research studies is 10,000 or more samples. However, do not be a slave to any particular number. The best approach is to understand what you are trying to estimate and to report not only point estimates but also standard errors and/or confidence intervals.

If you are only interested in the Monte Carlo estimate of the mean, then you can often use a relatively small number of samples. You can quantify this statement by looking at the Monte Carlo standard error, which is a measure of how close the mean of the sampling distribution is to the unknown population parameter. The Monte Carlo standard error is of the form $\sigma/\sqrt{m}$ where m is the number of samples that you generate (Ross 2006). This implies that to halve the Monte Carlo standard error, you need to quadruple the number of samples in the simulation.

Increasing the number of samples is one technique to reduce the Monte Carlo standard error, but other methods also exist. These so-called *variance reduction techniques* are described in Ross (2006, Ch. 8), Ripley (1987, Ch. 5), and Jones, Malllardet, and Robinson (2009, Ch. 20). There are two difficulties with using these techniques. First, they sometimes require considerable ingenuity because you need to construct certain auxiliary random variables with special properties. Second, they make programming the simulation more complicated because there is more "bookkeeping" in order to keep track of the auxiliary variables. In other words, these techniques enable you to use fewer simulations to obtain better estimates, but at the cost of a more complicated program.

6.4   Writing Efficient Simulations

The previous sections described several efficient simulation techniques. This section summarizes those techniques and describes additional techniques for efficiency.

6.4.1   The Basic Structure of Efficient Simulations in SAS

Recall that there are two basic techniques for simulating and analyzing data: the BY-group technique and the in-memory technique.

When you use the BY-group technique, do the following:

- Identify each sample by a unique value of the `SampleID` variable. Use a BY statement to compute statistics for each BY group.

- Suppress all output during the BY group analysis. Many procedures have a NOPRINT option in the PROC statement. Otherwise, use the method described in Section 6.4.2 to suppress ODS output.

When you use the SAS/IML in-memory technique, do the following:

- Use the J function to allocate a vector (or matrix) to store the simulated data before you call the RANDGEN subroutine. This enables you to generate an entire sample (or even multiple samples) with a single call to the RANDGEN subroutine. Do not generate one random value at a time.

- When possible, compute statistics for all samples with a single call. For example, the MEAN function can compute the means of all columns in a matrix. The subscript reduction operator `x[,:]` computes the means of all rows in the `x` matrix.

- If you are analyzing multivariate data, then it is often convenient to generate the data in a DO loop. At each iteration, generate a single sample and compute the statistic on that sample. This approach is described further in Chapter 8, "Simulating Data from Basic Multivariate Distributions."

6.4.2   How to Suppress ODS Output and Graphics

This section assumes that you are familiar with the Output Delivery System (ODS), as described in Section 3.5. ODS enables you to select, exclude, and output tables and graphics.

When you use a SAS procedure to compute statistics for each BY group, you should create an output data set that contains the statistics, as described in Section 4.4.1. Because you are writing the statistics to a data set, you do not need to display any output from the procedure.There are two ways to suppress procedure output: by using a NOPRINT option and by using the ODS EXCLUDE ALL statement and related ODS statements.

About 50 SAS/STAT procedures support the NOPRINT option in the PROC statement. When you specify the NOPRINT option, ODS is temporarily disabled while the procedure runs. This prevents SAS from displaying tables and graphs that would otherwise be produced for each BY group. For a simulation that computes statistics for thousands of BY groups, suppressing the display of tables results in a substantial savings of time. (But, the *SAS/STAT User's Guide* states, "However, there are a few procedures that for historical reasons still might produce some output even when NOPRINT is specified.")

The NOPRINT option is ideal for writing statistics to a data set by using procedure syntax such as the OUTPUT statement and other "OUT" options, such as the OUTP= option in PROC CORR, the OUT= option in PROC FREQ, the OUTEST= option in PROC REG, and the OUTSTAT= option in PROC GLM.

However, sometimes the statistic of interest is available only in an ODS table. In these cases, you cannot use the NOPRINT option because it suppresses all ODS tables, including the one that contains the statistic of interest. In these cases, use ODS to prevent tables and graphics from displaying on your computer monitor, but create an output data set by using the ODS OUTPUT statement.

For example, you can use the following technique to turn off ODS output. Prior to calling the procedure, execute the following statements:

```
/* suppress output to ODS destinations */
ods graphics off;
ods exclude all;
ods noresults;
```

The first statement turns off ODS graphics. Technically, you only need to use this statement prior to calling a procedure (such as the REG procedure) that produces ODS graphics by default, but there is no harm in unconditionally turning off ODS graphics. The second statement excludes all tables from open destinations such as HTML or LISTING. The third statement prevents ODS from making entries in the ODS Results window, which is shown in Figure 6.2.

Figure 6.2 The ODS Results Window

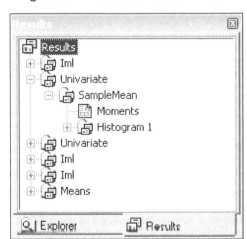

You can now run a SAS procedure without seeing any output displayed, and you can use the ODS OUTPUT statement to save a table of statistics to an output data set. After you compute the statistics for each BY group, re-enable ODS output by using the following statements:

```
ods graphics on;
ods exclude none;
ods results;
```

Because these sequences of commands are used so frequently in simulation studies, it is convenient to package them into SAS macros:

```
%macro ODSOff;                     /* Call prior to BY-group processing */
ods graphics off;
ods exclude all;
ods noresults;
%mend;
```

```
%macro ODSOn;                    /* Call after BY-group processing    */
ods graphics on;
ods exclude none;
ods results;
%mend;
```

With these definitions, you can easily disable ODS output temporarily while computing the statistics for each BY group. For example, the following statements write descriptive statistics to a data set named Desc, but suppresses the display of tables or graphs. The SimNormal data was created in Section 4.4.4.

```
%ODSOff
proc means data=SimNormal;
   by SampleID;
   var x;
   ods output Summary=Desc;
run;
%ODSOn
```

Sometimes it is convenient to also suppress SAS notes during a simulation. This is covered in the next section.

6.4.3   How to Suppress Notes to the SAS Log

Some SAS procedures write a note to the SAS log as part of their normal operation. For example, procedures that use maximum likelihood estimation write a note for each BY group that reports whether the numerical optimization succeeded for that BY group, as shown in the following example:

```
NOTE: Convergence criterion (GCONV=1E-8) satisfied.
NOTE: The above message was for the following BY group:
      SampleID=3
```

Not only do these messages take time to print, but they can also fill up the SAS log, which (when you are running SAS software interactively) results in a dialog box that gives you the option to clear the log. It is therefore desirable to deal with these messages in one of two ways:

- If you are confident that the notes are uninformative, then you can suppress notes by executing the following command:

  ```
  options nonotes;
  ```

  After the procedure runs, you can re-enable notes:

  ```
  options notes;
  ```

- If you are concerned that some of the notes might be important and you want the opportunity to review the SAS log after the simulation is complete, then you can redirect the SAS log to a file by using the PRINTTO procedure, as follows:

```
proc printto log='name-of-log-file' new;
run;
```

After the simulation completes, you can check the log file, reset the log, and restore the output destinations to their default values:

```
proc printto;                      /* no options ==> restore defaults */
run;
```

6.4.4   Caution: Avoid Macro Loops

Many SAS users attempt to run a simulation by using a macro loop instead of using the template presented in Section 4.4.1. Do not do this. Novikov (2003) compares times for the two methods and concludes (for his application) that the macro-loop technique is 80–100 times slower than the BY-group technique.

To be concrete, the following program is an example of macro code that computes the same quantities as in Section 4.4.2. You should avoid writing programs like this:

```
/************************************/
/* DO NOT USE THIS CODE: INEFFICIENT */
/************************************/
%macro Simulate(N, NumSamples);
options nonotes;                        /* turn off notes to log    */
proc datasets nolist;
   delete OutStats;                     /* delete data if it exists  */
run;

%do i = 1 %to &NumSamples;
   data Temp;                           /* create one sample         */
   call streaminit(0);
   do i = 1 to &N;
      x = rand("Uniform");
      output;
   end;
   run;

   proc means data=Temp noprint;        /* compute one statistic     */
      var x;
      output out=Out mean=SampleMean;
   run;

   proc append base=OutStats data=Out;          /* accumulate stats */
   run;
%end;
options notes;
%mend;

/* call macro to simulate data and compute ASD */
%Simulate(10, 100)                 /* means of 100 samples of size 10 */
```

This approach suffers from a low ratio of work to overhead. The DATA step and the MEANS procedure are called 100 times, but they generate or analyze only 10 observations in each call. This is inefficient because every time that SAS encounters a procedure call, it must parse the SAS code, open the data set, load data into memory, do the computation, close the data set, and exit the procedure. When a procedure computes complicated statistics on a large data set, these "overhead" costs are small relative to the computation performed by the procedure. However, for this example, the overhead costs are large relative to the computational work. For an example of a hybrid approach that uses macro and BY-group processing together, see the next section.

The macro approach also makes it difficult to reproduce the simulation. Notice that zero is used as the seed value, which means that the random number stream is initialized by using the system clock. In this example, you cannot call the STREAMINIT subroutine with a nonzero seed because then every sample data set would contain exactly the same data. An alternative to using a zero seed is to use the macro variable `&i` as the seed value.

Caution: If you do not turn off the NOTES option, then the performance of the `%Simulate` macro will be even worse. If the number of simulation loops is sufficiently large, then you might fill the SAS log with irrelevant text.

Exercise 6.1: Compare the run time for the `%Simulate` macro with the run time for the BY-group technique in Section 4.4.2. Because the NONOTES option is used, you might want to use PROC PRINT to print something when the macro completes.

6.4.5 When Macros Are Useful

The point of the previous section is that SAS simulations run faster when you use a small number of DATA steps and procedure calls to generate and analyze a large amount of data. The BY-group approach is efficient because a single call to a SAS procedure results in computing thousands of statistics.

This does not imply that macros are "bad." After all, a macro merely generates SAS code. However, if you use macros naively, then you might generate inefficient SAS code.

For huge simulation studies with a large number of design parameters, macros can be useful. Provided that a macro generates an efficient SAS program that uses BY-groups to read and process a lot of data, it is perfectly acceptable to encapsulate your simulation into a macro. In fact, for huge simulation studies, it makes sense to run the simulation as a series of smaller sub-studies. For example, for the simulation study in Section 6.2, you might write a macro that takes parameters for the sample size, the distribution of the error term, and the design of the independent variable. You could then call the macro $3 \times 5 \times 3$ times. Each call creates results that are sufficient to create a graph similar to Figure 6.1.

For huge simulation studies, encapsulating the program into a macro can provide several benefits:

- Efficiency: Novikov and Oberman (2007) noted that if you generate a single huge SAS data set that contains millions of samples, then the performance might be worse than running k analyses, each on $1/k$ samples. Novikov and Oberman suggest using values of k in the range of 20–50. The optimal value depends on the analysis, as well as performance characteristics of your computer, but the key point to remember is that running one huge simulation might take longer than several smaller (but still "large") simulations.

- Robustness: If you break up the study into several sub-studies, then you protect against catastrophic errors such as loss of power or a bug in your program that manifests itself just before the simulation completes. If you break the study into smaller sub-studies, then you will not lose all your results if something catastrophic occurs near the end of the simulation.

- Distributed processing: If you are conducting a study that involves a large number of independent parameter values and you have access to multiple computers that each have SAS software, then you can distribute the computations by running a subset of parameter values on each computer. Some researchers call simulation studies *embarrassingly parallel* because it is so easy to distribute the computations. You can distribute these computations manually, or you can use a tool such as SAS Grid Manager.

6.4.6   Profiling a SAS/IML Simulation

Some simulations take a long time to run. Consequently, it is essential to know how to *profile* a program, which means estimating the performance of various portions of the program. Wicklin (2010, p. 371) defines the *performance* of a program as "the time required for it to run on typical data, and also how that time changes with the size or characteristics of the data."

Suppose that part of your simulation involves finding the eigenvalues of a large matrix, and that the size of the matrix varies with the number of variables in the simulation. The following SAS/IML program generates random symmetric $n \times n$ matrices for various values of n, and times how long it takes to compute the eigenvalues:

```
proc iml;
size = do(500, 2000, 250);    /* 500, 1000, ..., 2000              */
time = j(1, ncol(size));       /* allocate vector for results       */
call randseed(12345);
do i = 1 to ncol(size);
   n = size[i];
   r = j(n*(n+1)/2, 1);        /* generate lower triangular elements */
   call randgen(r, "uniform");
   A = sqrvech(r);             /* create symmetric matrix           */

   t0 = time();
   evals = eigval(A);
   time[i] = time()-t0;        /* elapsed time for computation       */
end;
```

The TIME function returns the time of day as the number of seconds after midnight. The TIME function is called once prior to the eigenvalue computation and then again after the eigenvalue computation. The difference between those times is the elapsed time in seconds.

You can write the elapsed times to a SAS data set and use PROC SGPLOT to visualize the performance as a function of the size of the matrix. See Figure 6.3.

```
create eigen var {"Size" "Time"}; append; close;
quit;
```

```
proc sgplot data=eigen;
   title "Performance of Eigenvalue Computation";
   series x=Size y=Time / markers;
   yaxis grid label="Time to Compute Eigenvalues (s)";
   xaxis grid label="Size of Matrix";
run;
```

Figure 6.3 Performance of a Simple Computation

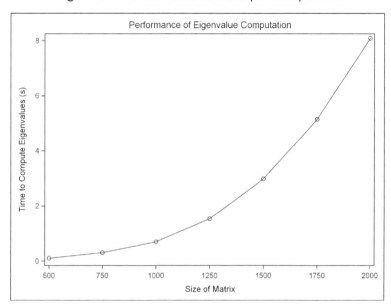

Figure 6.3 shows that the computation time increases nonlinearly with the size of the matrix. The time required for a 500 × 500 matrix is a fraction of a second. For a 2000 × 2000 matrix, the computation requires about eight seconds.

You can use this technique to time various components of your simulation. You can then focus on optimizing the program statements that are consuming the most time.

6.4.7   Tips for Shortening Simulation Times

This section collects some general suggestions for making your simulation run faster and more efficiently.

The following tips apply to SAS/IML programs:

- *Vectorize* computations. This means that you should write a relatively small number of statements and function calls, each of which performs a lot of work. For example, avoid loops over rows or elements of matrices. Instead, use matrix and vector computations.

- Allocate space for a matrix of random numbers, and then generate all of the random numbers with a single call to the RANDGEN subroutine.

- *Profile* the program to identify trouble spots. Profiling means timing different sections of a program to determine the run time for each section, as described in Section 6.4.6.

- Allocate result arrays outside of a loop; fill the array inside a loop. *Never* iteratively append to an array when performance is crucial.

- Optimize the performance of the statements in the innermost loop. These statements are executed most often.

In addition, there are some general performance tips that apply to simulation studies that involve parameters. For example, Section 5.4.2 uses a grid of uniformly spaced values to estimate the power of the *t* test. Similarly, Section 16.10 uses an evenly spaced grid of parameters to explore how skewness and kurtosis affect the coverage probability of a confidence interval. In studies such as these, a simulation is run for every parameter value, which means that in a straightforward implementation the total time is directly proportional to number of parameter values. Consequently, keep in mind the following tips:

- Reduce the size of your parameter grid. A 20×20 grid of parameters will take four times longer than a 10×10 grid. Use coarse grids to get a broad overview of the simulation results. Only refine the grid when the response function changes faster than the resolution of the grid.

- If you need to refine the grid, then consider local refinement rather than a global refinement. There is no reason to use a large number of grid points in regions where the response function is not changing much.

- Distribute independent computations. A 10×10 grid of parameter values corresponds to 100 independent simulations. If you have access to four computers, then consider running 25 simulations on each computer and collating the results.

- Estimate the run time of a simulation *before* you run it. If you plan to run 100,000 simulations, then start by running 100 simulations and time how long it takes. The final time will be about 1,000 times longer. This will let you know whether you should run the simulation while you go to lunch, or whether you should run it overnight.

- Remember that the number of points in a grid grows geometrically with the dimension. If you have two parameters, each with 10 possible values, then you have to run 10^2 simulations. If you attempt to use five parameters, and each has 10 possible values, then you have to run 10^5 simulations. Perhaps for some parameters you can consider only three values: small, medium, and large. Or perhaps the problem is symmetric in one of the parameters, and you do not need to consider using negative values. Or perhaps instead of using a complete factorial design you should consider using an incomplete factorial design.

- Respect the size of a billion. Due to advances in computing power, some people like to joke that "a billion is the new million," by which they mean that today you can run billions of computations in the same time that it once took to run millions of computations. That is true, but remember that it used to take a long time to run millions of computations.

It is a good idea to figure out how long it takes to generate a billion random normal variates on your computer. You can use this length of a time as a benchmark for running future simulations.

Exercise 6.2: Run a DATA step that generates one million samples, each containing 1,000 random normal observations. Check the SAS log to determine how long it takes to create that simulated data.

Exercise 6.3: Run a SAS/IML program that generates a vector of 1,000 random normal observations. Use the TIME function (see Section 6.4.6) to time how long it takes to fill the vector one million times within a DO loop. (You probably cannot generate all of these values in a single matrix. A billion doubles requires 8 GB of RAM, and SAS/IML 9.3 cannot allocate matrices larger than 2 GB.)

6.5   Disadvantages of Simulation

Simulation is a powerful technique, but it has limitations, which include difficulty in generalizing the results, difficulty in organizing the results, and difficulty in applying the results to real data.

A common criticism of simulation studies is that they cannot be generalized beyond the specific population models and the parameter values in the study. Furthermore, real data are rarely distributed according to *any* "named" distribution, so how well do the results generalize to real data?

As an example, suppose that you use simulation to reveal how deviations from normality affect the performance of a statistical test. In your study, you might simulate data from a *t* distribution and a lognormal distribution as two instances of nonnormal distributions. If the test performs well on the simulated nonnormal data, then you might want to conclude that the statistic is robust to departures from normality.

However, can you justify generalizing this result to other nonnormal distributions? Probably not. Do your conclusions hold for an exponential distribution or, more generally, for gamma distributions? Simulations provide *evidence* that something is true, but you need to be careful not to extrapolate beyond the cases that were simulated.

Because of the previous shortcoming, you might be tempted to run a huge simulation study that incorporates a wide range of nonnormal distributions, such as the exponential, gamma, and beta families. You might feel a need to run a simulation for every nonnormal distribution under the sun.

If you were to give in to these temptations, then how would you organize the results? Would you create pages of tables that list results for every distribution and every parameter value? Some researchers do. However, the fact that the study is not well designed makes it difficult to present the results in an organized manner.

Before you start simulating data, think about what you are trying to demonstrate. Perhaps "show this statistic works well for nonnormal data" is too nebulous a goal. Perhaps a more attainable goal is to study how the test behaves for populations with a wide range of skewness and kurtosis.

With this more modest goal, you can design a better simulation study. As shown in Chapter 16, you can sample from distributions whose skewness and kurtosis values lie on a regularly spaced grid (see Figure 16.16). With this design, you can display the results in an organized graph.

In spite of its limitations, simulation remains a powerful technique that belongs in the toolbox of every statistical programmer. Data simulation enables you to compare two or more statistical techniques by studying their performance on common data with known properties. Most importantly, data simulation enables you to understand the sampling distribution of statistics without resorting to asymptotic theory or overly restrictive assumptions about the distribution of data.

6.6 References

Gentle, J. E. (2009), *Computational Statistics*, New York: Springer-Verlag.

Jones, O., Maillardet, R., and Robinson, A. (2009), *Introduction to Scientific Programming and Simulation Using R*, Boca Raton, FL: Chapman & Hall/CRC.

Novikov, I. (2003), "A Remark on Efficient Simulations in SAS," *Journal of the Royal Statistical Society, Series D*, 52, 83–86.
URL http://www.jstor.org/stable/4128171

Novikov, I. and Oberman, B. (2007), "Optimization of Large Simulations Using Statistical Software," *Computational Statistics and Data Analysis*, 51, 2747–2752.

Ripley, B. D. (1987), *Stochastic Simulation*, New York: John Wiley & Sons.

Ross, S. M. (2006), *Simulation*, 4th Edition, Orlando, FL: Academic Press.

Wicklin, R. (2010), *Statistical Programming with SAS/IML Software*, Cary, NC: SAS Institute Inc.

Part III

Advanced Simulation Techniques

Chapter 7
Advanced Simulation of Univariate Data

Contents

7.1 Overview of Advanced Univariate Simulation

Chapter 2, "Simulating Data from Common Univariate Distributions," describes how to simulate data from a variety of discrete and continuous "named" distributions, such as the binomial, geometric, normal, and exponential distributions. This chapter explains how to simulate data from less commonly encountered univariate distributions and distributions with more complex structures.

7.2 Adding Location and Scale Parameters

You can add a location and a scale parameter to any distribution without changing its shape. If F is the (cumulative) distribution function of a random variable X, then the random variable

$Y = \theta + \sigma X$ for $\sigma > 0$ has the distribution function $F((x - \theta)/\sigma)$. If X has density function f, then Y has density $(1/\sigma) f((x - \theta)/\sigma)$ (Devroye 1986, p. 12). The parameter θ is a *location parameter*, whereas σ is a *scale parameter*. Parameters that are invariant under translation and scaling are called *shape parameters*.

In SAS 9.3, some SAS functions do not support a location or scale parameter for a distribution. For example, a general form for the density of the exponential distribution is $f(x; \theta, \sigma) = (1/\sigma) \exp(-(x - \theta)/\sigma)$ for $x > \theta$, where θ is a location parameter and $\sigma > 0$ is a scale parameter. This is the form supported by the UNIVARIATE procedure. However, the exponential distribution is parameterized in three other ways in SAS software:

- The RAND function does not support location or scale. The RAND function uses the standardized density $f(x; 0, 1) = e^{-x}$.

- The PDF, CDF, and QUANTILE functions do not support a location parameter. The exponential density used by these functions is $f(x; 0, \sigma) = (1/\sigma) \exp(-x/\sigma)$.

- The MCMC procedure supports the density $f(x; 0, 1/\lambda) = \lambda e^{-\lambda x}$, where $\lambda > 0$ is a rate parameter.

Fortunately, you can include location and scale parameters even when a function does not explicitly support them:

- To simulate a random variate Y with location parameter θ and scale parameter σ, use the RAND function to generate the standard random variate X and form $Y = \theta + \sigma X$.

- To compute a PDF that incorporates location and scale, call the PDF function as `(1/sigma)* PDF("DistribName", (x-theta)/sigma)`.

- To compute a CDF that incorporates location and scale, call the CDF function as `CDF("DistribName", (x-theta)/sigma)`.

- To compute a quantile that incorporates location and scale, call the QUANTILE function as `x = theta + sigma*QUANTILE("DistribName", p)`, where p is the probability such that $P(X \leq x) = p$.

- To incorporate a rate parameter, define the scale parameter as $\sigma = 1/\lambda$.

SAS software provides about two dozen "named" distributions as listed in Section 2.7. For a few distributions, additional parameters are sometimes needed. The following list shows how to generate random variates with parameters that are not supported by the RAND function. The symbol $\sim$ means "is distributed according to." For example, $X \sim N(0, 1)$ means that the random variable X follows a normal distribution with mean 0 and unit standard deviation.

Exponential: The RAND function generates $E \sim \text{Exp}(1)$. The random variable σE is an exponential random variable with scale parameter σ. The random variable E/λ is an exponential random variable with rate parameter λ.

Gamma: The RAND function generates $G \sim \text{Gamma}(\alpha)$. The random variable σG is a gamma random variable with scale parameter σ.

Lognormal: The RAND function generates $N \sim N(\mu, \sigma)$. The random variable $\exp(N)$ is lognormally distributed with parameters μ and σ.

Uniform: The RAND function generates $U \sim U(0, 1)$. The random variable $a + (b - a)U$ is uniformly distributed on the interval (a, b).

For the lognormal distribution, notice that the location and scale parameters are added *before* the exponential transformation is applied.

Exercise 7.1: Simulate 1,000 exponential variates with scale parameter 2. Overlay the corresponding PDF. Compute the sample median. Is it close to the theoretical value, which is $2 \ln(2) \approx 1.386$?

Exercise 7.2: Simulate 1,000 lognormal variates by simulating $Y \sim N(1.5, 3)$ and applying the transformation $X = \exp(Y)$. Use PROC UNIVARIATE to verify that X is lognormally distributed with parameters 1.5 and 3.

7.3 Simulating from Less Common Univariate Distributions

You can often use random variates from simple distributions (in particular, the normal, exponential, and uniform distributions) to simulate data from more complicated distributions.

To illustrate this approach, this section shows how to simulate data from the Gumbel, inverse Gaussian, Pareto, generalized Pareto, power function, Rayleigh, Johnson S_B, and Johnson S_U distributions. Most of these distributions are supported by PROC UNIVARIATE, but they are not supported by the RAND function in SAS 9.3.

To simulate from other distributions, see the algorithms in Devroye (1986) and the "Details" section of the PROC MCMC documentation in the *SAS/STAT User's Guide*.

7.3.1 The Gumbel Distribution

The Gumbel distribution is sometimes called the Type I extreme value distribution (Johnson, Kotz, and Balakrishnan 1994, p. 686). It is the limiting distribution of the greatest value among n independent random variables, as $n \to \infty$. The density function for the Gumbel distribution with location μ and scale σ is

$$f(x) = e^{-(x-\mu)/\sigma} \exp\left(-e^{-(x-\mu)/\sigma}\right)$$

To simulate data from a Gumbel distribution, let $E \sim \text{Exp}(1)$ and form $X = \mu + \sigma(-\log(E))$ (Devroye 1986, p. 414), as follows:

```
/* Simulate from Gumbel distribution (Devroye, p. 414) */
E = rand("Exponential");
x = mu + sigma*(-log(E));
```

7.3.2   The Inverse Gaussian Distribution

The inverse Gaussian distribution (also called the *Wald distribution*) describes the first-passage time of Brownian motion with positive drift. The UNIVARIATE procedure supports fitting an inverse Gaussian distribution with a location parameter, $\mu > 0$, and a shape parameter, $\lambda > 0$. The density function for $x > 0$ is

$$f(x) = \left(\frac{\lambda}{2\pi x^3}\right)^{1/2} \exp\left(\frac{-\lambda(x-\mu)^2}{2\mu^2 x}\right)$$

The RANDGEN function in SAS/IML 12.1 software supports simulating data from the inverse Gaussian distribution. However, the RAND function in SAS 9.3 software does not. For completeness, the following DATA step simulates inverse Gaussian data. The algorithm is from Devroye (1986, p. 149), which shows how to use normal and uniform variates to simulate data from an inverse Gaussian distribution.

```
/* Simulate from inverse Gaussian (Devroye, p. 149) */
data InvGauss(keep= X);
mu = 1.5;                              /* mu > 0     */
lambda = 2;                            /* lambda > 0 */
c = mu/(2 * lambda);
call streaminit(1);
do i = 1 to 1000;
   muY = mu * rand("Normal")**2;       /* or mu*rand("ChiSquare", 1) */
   X = mu + c*muY - c*sqrt(4*lambda*muY + muY**2);
   /* return X with probability mu/(mu+X); otherwise mu**2/Y */
   if rand("Uniform") > mu/(mu+X) then /* or rand("Bern", X/(mu+X)) */
      X = mu*mu/X;
   output;
end;
run;
```

7.3.3   The Pareto Distribution

The RANDGEN function in SAS/IML 12.1 software supports simulating data from the Pareto distribution. The Pareto distribution with shape parameter $a > 0$ and scale parameter is $k > 0$ has the following density function for $x > k$:

$$f(x) = \frac{a}{k}\left(\frac{k}{x}\right)^{a+1}$$

In SAS 9.3, the RAND function does not support the Pareto distribution. The CDF for the Pareto distribution is $F(x) = 1 - (k/x)^a$; therefore, you can use the inverse CDF method (see Section 7.4) to simulate random variates (Devroye 1986, p. 29), as shown in the following DATA step:

```
/* Simulate from Pareto (Devroye, p. 29) */
data Pareto(keep= X);
a = 4;                        /* alpha > 0                              */
k = 1.5;                      /* scale > 0 determines lower limit for x */
call streaminit(1);
do i = 1 to 1000;
   U = rand("Uniform");
   X = k / U**(1/a);
   output;
end;
run;
```

The algorithm uses the fact that if $U \sim U(0, 1)$, then also $(1 - U) \sim U(0, 1)$.

7.3.4   The Generalized Pareto Distribution

The generalized Pareto distribution has the following cumulative distribution:

$$F(z; \alpha) = 1 - (1 - \alpha z)^{1/\alpha}$$

where α is the shape parameter and $z = (x - \theta)/\sigma$ is a standard variable. (In spite of its name, there is not a simple relationship between these parameters and the parameters of the usual Pareto distribution.) You can use the inverse CDF algorithm (see Section 7.4) to simulate random variates. In particular, let $U \sim U(0, 1)$ be a uniform variate. Then $Z = -(1/\alpha)(U^{\alpha} - 1)$ is a standardized generalized Pareto variate. Equivalently, the following statements simulate a generalized Pareto variate:

```
/* Generalized Pareto(threshold=theta, scale=sigma, shape=alpha) */
U = rand("Uniform");
X = theta - sigma/alpha * (U**alpha-1);
```

The algorithm uses the fact that if $U \sim U(0, 1)$, then also $(1 - U) \sim U(0, 1)$.

7.3.5   The Power Function Distribution

The power function distribution with threshold parameter θ, scale parameter σ, and shape parameter α has the following density function for $\theta < x < \theta + \sigma$:

$$f(x) = \frac{\alpha}{\sigma} \left(\frac{x - \theta}{\sigma} \right)^{\alpha - 1}$$

If $E \sim \text{Exp}(1)$, then $Z = (1 - \exp(-E))^{1/\alpha}$ follows a standard power function distribution (Devroye 1986, p. 262). In the general case, you can simulate a random variate that follows the power function distribution by using the following statements:

```
/* power function(threshold=theta, scale=sigma, shape=alpha) */
E = rand("Exponential");
X = theta + sigma*(1-exp(-E))**(1/alpha);
```

7.3.6   The Rayleigh Distribution

The Rayleigh distribution with threshold parameter θ and scale parameter σ has the following density function for $x \geq \theta$:

$$f(x) = \frac{x - \theta}{\sigma^2} e^{-(x-\theta)^2/(2\sigma^2)}$$

The Rayleigh distribution is a special case of the Weibull distribution. A Rayleigh random variable with scale parameter σ is the same as a Weibull random variable with shape parameter 2 and scale parameter $\sqrt{2}\sigma$. Consequently, the following statement generates a random variate from the Rayleigh distribution:

```
/* Rayleigh(threshold=theta, scale=sigma) */
X = theta + rand("Weibull", 2, sqrt(2)*sigma);
```

Alternatively, you can use the inverse CDF method (Devroye 1986, p. 29).

7.3.7   The Johnson S_B Distribution

The Johnson system of distributions (Johnson 1949) is described in Section 16.7 and in the PROC UNIVARIATE documentation. The Johnson system is a flexible way to model data (Bowman and Shenton 1983).

Johnson's system consists of three distributions that are defined by transformations to normality: the S_B distribution (which is bounded), the S_L distribution (which is the lognormal distribution), and the S_U distribution (which is unbounded). Technically, the Johnson family also contains the normal distribution. The distributions can be defined in terms of four parameters: two shape parameters γ and δ, a threshold parameter (θ), and a scale parameter (σ). The parameter σ is positive, and by convention δ is also positive. (Using a negative δ leads to distributions with negative skewness.)

A random variable X follows a Johnson distribution if

$$Z = \gamma + \delta g_i(X; \theta, \sigma)$$

is a standard normal random variable. Johnson's idea was to choose functions $g_i, i = 1, 2, 3$ so that the three distributions encompass all possible values of skewness and kurtosis, and consequently describe a wide range of shapes of distributions.

For the S_B distribution, the normalizing transformation is

$$g_1(x; \theta, \sigma) = \log\left(\frac{x - \theta}{\sigma + \theta - x}\right)$$

The S_B distribution is not built into the RAND function. Nevertheless, you can use the definition of the S_B distribution to simulate the data, as shown in the following program:

```
/* Johnson SB(threshold=theta, scale=sigma, shape=delta, shape=gamma) */
data SB(keep= X);
call streaminit(1);
theta = -0.6;    scale = 18;    delta = 1.7;    gamma = 2;
```

```
do i = 1 to 1000;
   Y = (rand("Normal")-gamma) / delta;
   expY = exp(Y);
   /* if theta=0 and sigma=1, then X = logistic(Y) */
   X = ( sigma*expY + theta*(expY + 1) ) / (expY + 1);
   output;
end;
run;
```

This technique—simulate from the normal distribution and then apply a transformation in order to get a new distribution with desirable properties—is a fundamental technique in simulation. It is used again in Chapter 16 to construct the Fleishman distribution, which has a specified skewness and kurtosis. It is used often in Chapter 8 to construct multivariate distributions that have specified correlations and marginal distributions.

For completeness, the normalizing transformation for the S_L distribution is

$$g_2(x; \theta, \sigma) = \log\left(\frac{x - \theta}{\sigma}\right)$$

7.3.8   The Johnson S_U Distribution

For the Johnson S_U distribution, the normalizing transformation is

$$g_3(x; \theta, \sigma) = \sinh^{-1}\left(\frac{x - \theta}{\sigma}\right)$$

If X follows a $S_U(\theta, \sigma, \gamma, \delta)$ distribution, then $Z = \gamma + \delta \sinh^{-1}((X - \theta)/\sigma)$ is normally distributed. This means that you can generate a random variate from a Johnson S_U distribution by transforming a normal variate. By inverting the transformation, you can simulate data from the Johnson S_U distribution, as follows:

```
/* Johnson SU(threshold=theta, scale=sigma, shape=delta, shape=gamma) */
data SU(keep= X);
call streaminit(1);
theta = 1;  sigma = 5;   delta = 1.5;  gamma = -1;
do i = 1 to 10000;
   Y = (rand("Normal")-gamma) / delta;
   X = theta + sigma * sinh(Y);
   output;
end;
run;
```

You can use PROC UNIVARIATE to verify that the simulated data are from the specified S_U distribution. Figure 7.1 shows the parameter estimates. The estimates are somewhat close to the parameters, but there appears to be substantial variability in the estimates.

```
proc univariate data=SU;
   histogram x / su noplot;
   ods select ParameterEstimates;
run;
```

Figure 7.1 Parameter Estimates for Simulated S_U Data

The UNIVARIATE Procedure
Fitted SU Distribution for X

Parameters for Johnson SU Distribution		
Parameter	Symbol	Estimate
Location	Theta	0.797372
Scale	Sigma	5.122924
Shape	Delta	1.510779
Shape	Gamma	-1.03154
Mean		5.4982
Std Dev		5.526751
Skewness		1.737434
Kurtosis		8.323503
Mode		3.382778

Exercise 7.3: Simulate 10,000 observations from each of the following distributions. Use PROC UNIVARIATE to verify that the parameter estimates are close to the parameter values.

1. the Gumbel distribution with $\mu = 1$ and $\sigma = 2$

2. the inverse Gaussian distribution with $\mu = 1.5$ and $\sigma = 2$

3. the generalized Pareto distribution with $\theta = 0$, $\sigma = 2$, and $\alpha = -0.15$

4. the power function distribution with $\theta = 0$, $\sigma = 2$, and $\alpha = 0.15$

5. the Rayleigh distribution with $\theta = 1$ and $\sigma = 2$

Exercise 7.4: Simulate 1,000 samples, each of size 100, from the S_U distribution with $\theta = 1$, $\sigma = 5, \delta = 1.5$, and $\gamma = -1$. Use PROC UNIVARIATE to compute the corresponding parameter estimates. (Be sure to use OPTION NONOTES for this simulation.) Draw histograms of the parameter estimates. Which estimate has the largest variance? The smallest variance? Be aware that the four-parameter estimation might not succeed for every sample.

7.4 Inverse CDF Sampling

If you know the cumulative distribution function (CDF) of a probability distribution, then you can always generate a random sample from that distribution. However, this method of sampling can be computationally expensive unless you have a formula for the inverse CDF.

This sampling technique uses the fact that a continuous CDF, F, is a one-to-one mapping of the domain of the CDF into the interval $(0, 1)$. Therefore, if U is a random uniform variable on $(0, 1)$, then $X = F^{-1}(U)$ has the distribution F. For a proof, see Ross (2006, p. 62).

7.4.1   The Inverse Transformation Algorithm

To illustrate the inverse CDF sampling technique (also called the *inverse transformation algorithm*), consider sampling from a standard exponential distribution. The exponential distribution has probability density $f(x) = e^{-x}, x \geq 0$, and therefore the cumulative distribution is $F(x) = \int_0^x e^{-t}\, dt = 1 - e^{-x}$. This function can be explicitly inverted by setting $u = F(x)$ and solving for x. The inverse CDF is $x = F^{-1}(u) = -\log(1 - u)$.

The following DATA step generates random values from the exponential distribution by generating random uniform values from $U(0, 1)$ and applying the inverse CDF of the exponential distribution. The UNIVARIATE procedure is used to check that the data follow an exponential distribution.

```
/* Inverse CDF algorithm */
%let N = 100;                        /* size of sample */
data Exp(keep=x);
call streaminit(12345);
do i = 1 to &N;
   u = rand("Uniform");
   x = -log(1-u);
   output;
end;
run;

proc univariate data=Exp;
   histogram x / exponential(sigma=1) endpoints=0 to 6 by 0.5;
   cdfplot x / exponential(sigma=1);
   ods select GoodnessOfFit Histogram CDFPlot;
run;
```

Figure 7.2 Goodness-of-Fit Tests for Exponential Distribution

The UNIVARIATE Procedure
Fitted Exponential Distribution for x

Goodness-of-Fit Tests for Exponential Distribution				
Test		Statistic		p Value
Kolmogorov-Smirnov	D	0.10897832	Pr > D	0.179
Cramer-von Mises	W-Sq	0.44519963	Pr > W-Sq	0.057
Anderson-Darling	A-Sq	2.42878591	Pr > A-Sq	0.056

Figure 7.3   Histogram and Empirical CDF of Exponential Data

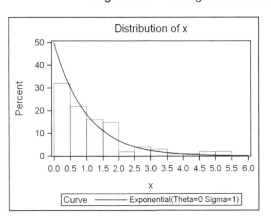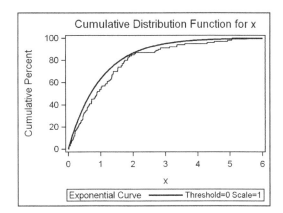

This technique is used by many statistical software packages to generate exponential random variates. Additional examples are shown in Chapter 5 of Ross (2006), which includes the current example.

Exercise 7.5: The function $F(x) = x^2$ defines a distribution on the interval $[0, 1]$. Use the inverse function to generate 100 random values from F. Use PROC UNIVARIATE to plot the empirical CDF, which should approximate F.

7.4.2   Root Finding and the Inverse Transformation Algorithm

Even if you cannot invert the CDF, you can still use the inverse CDF algorithm by numerically solving for the quantiles of the CDF. However, because this method is computationally expensive, you should use this technique only when direct methods are not available.

If F is a distribution function, then you can simulate values from F by doing the following:

1. Generate a random uniform value u in $(0, 1)$.

2. Use bisection or some other root-finding algorithm to find the value x such that $F(x) = u$.

3. Repeat these steps until you have generated N values of x.

The SAS/IML 12.1 language supports the FROOT function for finding the root of a univariate function. Prior to SAS/IML 12.1, you can use the Bisection module, which is defined in Appendix A.

To illustrate this method, define the distribution function $Q(x) = (x + x^3 + x^5)/3$ for x on the interval $[0, 1]$. You can test the bisection algorithm by defining the Func module to be the function $Q(x) - u$:

```
proc iml;
/* a quantile is a zero of the following function */
start Func(x) global(target);
   cdf = (x + x##3 + x##5)/3;
   return( cdf-target );
finish;
```

```
/* test bisection module */
target = 0.5;                    /* global variable used by Func module */
/* for SAS/IML 9.3 and before, use q = Bisection(0,1); */
q = froot("Func", {0 1});       /* SAS/IML 12.1                      */
```

The result of the computation is a quantile of the Q distribution. If you plug the quantile into Q, then you obtain the target value. Figure 7.4 shows that the median of the quintic distribution function is approximately 0.77.

```
p = (q + q##3 + q##5)/3;       /* check whether F(q) = target       */
print q p[label="CDF(q)"];
```

Figure 7.4 Median of Distribution

q	CDF(q)
0.7706141	0.5

To simulate values from Q, generate random uniform values as target values and use the root-finding algorithm to solve for the corresponding quantiles. Figure 7.5 shows the empirical density and ECDF for 100 random observations that are drawn from Q.

```
N = 100;
call randseed(12345);
u = j(N,1); x = j(N,1);
call randgen(u, "Uniform");              /* u ~ U(0,1)              */
do i = 1 to N;
   target = u[i];
   /* for SAS/IML 9.3 and before, use x[i] = Bisection(0,1); */
   x[i] = froot("Func", {0 1});          /* SAS/IML 12.1           */
end;
```

Figure 7.5 Histogram and Empirical CDF of Quintic Distribution

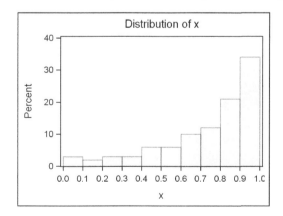
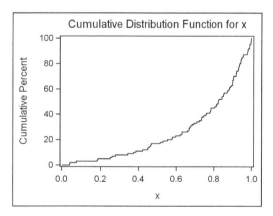

7.5 Finite Mixture Distributions

A mixture distribution is a population that contains subpopulations, which are also called *components*. Each subpopulation can be a different distribution, but they can also be the same distribution with different parameters. In addition to specifying the subpopulations, a mixture distribution requires that you specify *mixing probabilities*.

In general, a (finite) mixture distribution is composed of k components. If f_i is the PDF of the ith component, and π_i are the mixing probabilities, then the PDF of the mixture is $g(x) = \Sigma_{i=1}^{k} \pi_i f_i(x)$, where $\Sigma_{i=1}^{k} \pi_i = 1$. You can use the RAND function with the "Table" distribution to randomly select a subpopulation according to the mixing probabilities. For many choices of component densities, you can use the FMM procedure in SAS/STAT software to fit a finite mixture distribution to data.

7.5.1   Simulating from a Mixture Distribution

Suppose that you use a mixture distribution to model the time required to answer questions in a call center. From historic data, you subdivide calls into three types, each with its own distribution:

1. Easy questions: These questions can be answered in about three minutes. They account for about 50% of all calls. You decide to model this subpopulation by $N(3, 1)$.

2. Specialized questions: These questions require about eight minutes to answer. They account for about 30% of all calls. You decide to model this subpopulation by $N(8, 2)$.

3. Difficult questions: These questions require about 10 minutes to answer. They account for the remaining 20% of calls. You model this subpopulation by $N(10, 3)$.

The following DATA step uses the "Table" distribution (see Section 2.4.5) to randomly select a type of question (easy, specialized, or difficult) according to historical proportions. The time required to answer each question is then modeled as a random sample from the relevant subpopulation. Figure 7.6 shows the distribution of the simulated times overlaid with a kernel density estimate (KDE).

```
%let N = 100;                                      /* size of sample    */
data Calls(drop=i);
call streaminit(12345);
array prob [3] _temporary_ (0.5 0.3 0.2);
do i = 1 to &N;
   type = rand("Table", of prob[*]);              /* returns 1, 2, or 3 */
   if type=1 then      x = rand("Normal",  3, 1);
   else if type=2 then x = rand("Normal",  8, 2);
   else                x = rand("Normal", 10, 3);
   output;
end;
run;

proc univariate data=Calls;
   ods select Histogram;
   histogram x / vscale=proportion
   kernel(lower=0 c=SJPI);
run;
```

Figure 7.6 Sample from Mixture Distribution ($N = 100$)

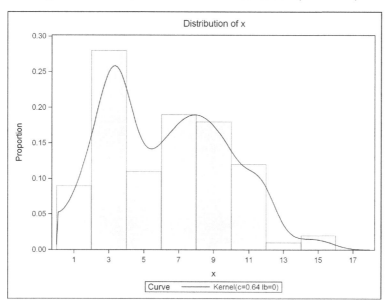

Figure 7.6 shows that the histogram and KDE have a major peak at $x = 3$, which corresponds to the mean of the most frequent type of question ("easy"). There is also a smaller peak at $x = 8$, which corresponds to the mean time for answering the "specialized" questions. The KDE also shows a small "bump" near $x = 10$, which corresponds to the "difficult" questions.

In order to simulate many samples, you can add an additional DO loop:

```
do SampleID = 1 to &NumSamples;          /* simulation loop */
   do i = 1 to &N;
   ...
   end;
end;
```

From a modeling perspective, a normal distribution is probably not a good model for the "easy" questions because the $N(3, 1)$ distribution can produce negative "times." The next exercise addresses this issue by truncating the mixture distribution at a positive value. (This is called a *truncated distribution*; see Section 7.7.)

Exercise 7.6: Assume that every call requires at least one minute to handle. Modify the DATA step in this section so that if x is less than 1, then it is replaced by the value 1. (This results in a mixture of a point mass and the truncated mixture distribution.) Simulate 1,000 samples from the distribution. Use PROC MEANS to compute the total time that is required to answer 100 calls. Use PROC UNIVARIATE to examine the distribution of the total times. Based on the simulation, predict how often the call center will need 11 or more hours to service 100 calls.

7.5.2   The Contaminated Normal Distribution

The contaminated normal distribution was introduced by Tukey (1960). The contaminated normal distribution is a specific instance of a two-component mixture distribution in which both components are normally distributed with a common mean. For example, a common contaminated normal model is to simulate values from an $N(0, 1)$ distribution with probability 0.9 and from an $N(0, 10)$ distribution with probability 0.1. This results in a distribution with heavier tails than normality. This is also a convenient way to generate data with outliers.

Because there are only two groups, you can use a Bernoulli random variable (instead of the more general "Table" distribution) to select the component from which each observation is drawn, as shown in the following program:

```
%let std = 10;                        /* magnitude of contamination */
%let N = 100;                         /* size of sample              */
data CN(keep=x);
call streaminit(12345);
do i = 1 to &N;
   if rand("Bernoulli", 0.1) then
      x = rand("Normal", 0, &std);
   else
      x = rand("Normal");
   output;
end;
run;


proc univariate data=CN;
   var x;
   histogram x / kernel vscale=proportion endpoints=-15 to 21 by 1;
   qqplot x;
run;
```

Figure 7.7   Histogram and Q-Q Plot of Contaminated Normal Sample

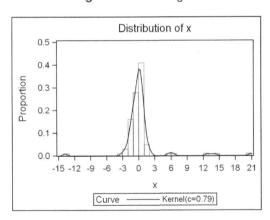 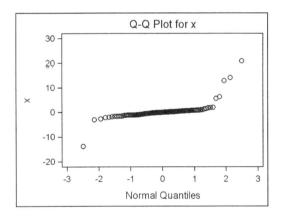

Johnson (1987, p. 55) states that "Many commonly used statistical methods perform abominably" with data that are generated from a contaminated normal distribution.

Exercise 7.7: It is possible to create a *continuous mixture distribution* by replacing a parameter with a random variable (Devroye 1986, p. 16). Simulate data from the distribution $N(\mu, 1)$, where $\mu \sim U(0, 1)$. Draw a histogram of the result. The mean of the simulated data should be close to $E(\mu) = 1/2$.

7.6 Simulating Survival Data

In medical and pharmaceutical studies, it is common to encounter time-to-event data. For example, suppose that a study contains 100 subjects. When a subject experiences the event of interest (death, remission, onset of disease,...), the time of the event is recorded.

In the simplest situation, the event occurs at a common constant rate, which is called the *hazard rate*. A constant rate is equivalent to the event times being exponentially distributed. You can use the %RANDEXP macro, as described in Section 2.5.3, to generate survival times. You can analyze the survival time by using the LIFETEST procedure, as follows. The results are shown in Figure 7.8.

```
/* sigma is scale parameter; use sigma=1/lambda for a rate parameter */
%macro RandExp(sigma);
    ((&sigma) * rand("Exponential"))
%mend;

data LifeData;
call streaminit(1);
do PatientID = 1 to 100;
    t = %RandExp(1/0.01);                    /* hazard rate = 0.01 */
    output;
end;
run;

proc lifetest data=LifeData;
    time t;
    ods select Quartiles Means;
run;
```

Figure 7.8 Mean and Quartiles of Survival Time

The LIFETEST Procedure

		Quartile Estimates		
			95% Confidence Interval	
Percent	Point Estimate	Transform	[Lower	Upper)
75	100.657	LOGLOG	89.226	119.829
50	64.512	LOGLOG	46.072	81.495
25	23.804	LOGLOG	15.241	37.199

Mean	Standard Error
84.663	8.990

The output in Figure 7.8 shows estimates for the mean survival time and for quartiles of the survival time. For comparison, the exponential distribution with scale parameter σ has quartiles given by

$Q_{25} = \sigma \log(4/3)$, $Q_{50} = \sigma \log(2)$, and $Q_{75} = \sigma \log(4)$. The mean of the exponential distribution is σ. For $\sigma = 100$, these quartiles are approximately $Q_{25} \approx 28.8$, $Q_{50} \approx 69.3$, $Q_{75} \approx 138.6$.

The second example introduces *censored* observations. If a subject completes the study without experiencing the event, then the event time for that subject is said to be censored. Similarly, patients who drop out of the study prior to experiencing the event are said to be censored. Suppose that you want to simulate the following situations:

- The hazard rate for each subject is 0.01 events per day.

- The rate at which subjects drop out of the study is 0.001 per day.

- The study lasts for 365 days.

In the following DATA step, the variable Censored is an indicator variable that records whether the recorded time is the time of the event (Censored= 0) or the time of censoring (Censored= 1). The LIFETEST procedure is used to analyze the results. The output is shown in Figure 7.9 and Figure 7.10.

```
data CensoredData(keep= PatientID t Censored);
call streaminit(1);
HazardRate = 0.01;          /* rate at which subject experiences event */
CensorRate = 0.001;         /* rate at which subject drops out         */
EndTime = 365;              /* end of study period                     */
do PatientID = 1 to 100;
   tEvent = %RandExp(1/HazardRate);
   c = %RandExp(1/CensorRate);
   t = min(tEvent, c, EndTime);
   Censored = (c < tEvent | tEvent > EndTime);
   output;
end;
run;

proc lifetest data=CensoredData plots=(survival(atrisk CL));
   time t*Censored(1);
   ods select Quartiles Means CensoredSummary SurvivalPlot;
run;
```

Figure 7.9 Summary Statistics for Estimates of Survival Time for Censored Data

The LIFETEST Procedure

Quartile Estimates				
		95% Confidence Interval		
Percent	Point Estimate	Transform	[Lower	Upper)
75	113.010	LOGLOG	91.696	148.567
50	63.599	LOGLOG	46.072	81.495
25	28.752	LOGLOG	17.113	40.600

Figure 7.9 *continued*

Mean	Standard Error
86.216	8.280

Note: The mean survival time and its standard error were underestimated because the largest observation was censored and the estimation was restricted to the largest event time.

Summary of the Number of Censored and Uncensored Values			
Total	Failed	Censored	Percent Censored
100	93	7	7.00

Figure 7.10 Survival Curve for Censored Data

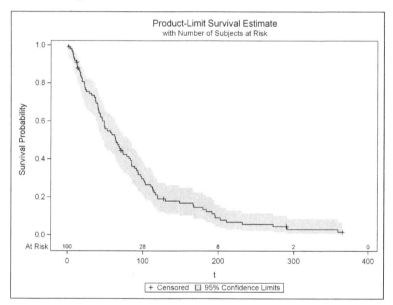

The analysis of censored data is sometimes called *survival analysis* because the event of interest is often death, but the event could also be relapse, remission, or recovery. Figure 7.10 shows the survival curve for the simulated patients. By Day 100, only 28 patients had not experienced the event. By Day 300, all but two patients had either dropped out of the study or had experienced the event.

In general, you can specify any distribution for the event times, as shown in the following exercise. See Section 12.4 for modeling survival times that include covariates.

Exercise 7.8: Simulate survival times that are lognormally distributed. For example, use `exp(rand("Normal", 4, 1))` as the time to event. First, assume no censoring and an arbitrarily long study. Next, assume a 365-day study and a constant rate of dropping out.

7.7   The Acceptance-Rejection Technique

The *rejection method* is a technique for simulating values from a distribution (called the *instrumental distribution*) that is subject to constraints. The idea is that you generate values from the instrumental distribution and then throw away (reject) any values that do not meet the constraints. It is also known as the *acceptance-rejection* technique. This section explores an instructive application of the rejection method: simulating data from a truncated normal distribution (Robert 1995).

You can truncate any distribution (Devroye 1986). If X is a random variable with instrumental distribution function F, then the truncated distribution on the interval $[a, b]$ is defined as

$$G(x) = \begin{cases} 0 & x < a \\ \frac{F(x)-F(a)}{F(b)-F(a)} & a \leq x \leq b \\ 1 & x > b \end{cases}$$

Here $-\infty \leq a < b \leq \infty$, with the convention that $F(-\infty) = 0$ and $F(\infty) = 1$.

In particular, a truncated standard normal distribution with left truncation point a has the cumulative distribution function $(\Phi(x) - \Phi(a))/(1 - \Phi(a))$, where Φ is the standard normal distribution function.

The obvious simulation algorithm for a truncated normal random variable is to simulate from a normal distribution and discard any value outside of the interval $[a, b]$. For example, the following DATA step simulates data from the truncated normal distribution with left truncation point $a = 0$:

```
%let N = 100;                          /* size of sample */
data TruncNormal(keep=x);
call streaminit(12345);
a = 0;
do i = 1 to &N;
   do until( x>=a );                   /* reject x < a    */
      x = rand("Normal");
   end;
   output;
end;
run;
```

In a similar manner, you can simulate from a doubly truncated normal distribution on the interval $[0, 2]$:

```
do until( (x>0 & x<2) );               /* x in (0, 2) */
   x = rand("Normal");
end;
```

This technique is appropriate when the truncation points are near the middle of the normal distribution. For truncation points that are in the tails of the normal distribution, see the accept-reject algorithm in Robert (1995). See also Devroye (1986, p. 380).

An alternative approach to the rejection technique is to use the inverse CDF technique. This technique is applicable for any truncated distribution G, but requires computing the inverse CDF, which can be expensive. To implement this technique, first generate $U \sim U(0, 1)$. Then the random variable $Y = G^{-1}(U) = F^{-1}(F(a) + U(F(b) - F(a)))$ is distributed according to G. For the left-truncated standard normal distribution, use the following SAS statements:

```
Phi_a = cdf("Normal", 0);            /* a = 0          */
Phi_b = 1;                           /* b = infinity */
do i = 1 to &N;
   u = rand("Uniform");
   x = quantile("Normal", Phi_a + u*(Phi_b - Phi_a));
   output;
end;
```

For implementing univariate accept-reject techniques, the DATA step has an advantage over the SAS/IML language. It can be a challenge to implement time-to-event and accept-reject algorithms in the SAS/IML language because you do not know ahead of time how many random values to simulate. An effective technique is to multiply the number of observations that you want by some large factor. For example, for the truncated normal distribution with a truncation point of zero, you can expect half of the random normal variates to be rejected. To be on the safe side, simulate more than twice as many normal variates as you need for the truncated normal distribution as shown in the following SAS/IML program:

```
proc iml;
call randseed(12345);
multiple = 2.5;                  /* choose value > 2           */
y = j(multiple * &N, 1);         /* allocate more than you need */
call randgen(y, "Normal");       /* y ~ N(0,1)                 */
idx = loc(y > 0);                /* acceptance step            */
x = y[idx];
x = x[1:&N];                     /* discard any extra observations */
```

A more statistical approach is to use the QUANTILE function to estimate the proportion of simulated data that are likely to be rejected during the simulation. You then sample enough variates from the instrumental distribution to be 99.9% certain that you will obtain at least N observations from the truncated distribution.

For the truncated normal distribution, $p = 0.5$ is the probability that a variate from the instrumental distribution is accepted. The event "accept the instrumental variate" is a Bernoulli random variable with probability of success p. The negative binomial is the distribution of the number of failures before N successes in a sequence of independent Bernoulli trials. You can compute F, the 99.9th percentile of the number of failures that you encounter before you obtain N successes. Therefore, the total number of trials that you need is $M = F + N$.

For the truncate normal distribution, the following SAS/IML statements compute these quantities. The computation indicates that you should generate 248 normal variates in order to be 99.9% certain that you will obtain at least 100 variates for the truncated normal distribution. See Figure 7.11.

```
p = 0.5;                  /* prob of accepting instrumental variate */
F = quantile("NegBin", 0.999, p, &N);
M = F + &N;               /* Num Trials = failures + successes      */
print M;
```

Figure 7.11 Number of Trials to Generate $N = 100$ Variates

M
248

Exercise 7.9: Create a histogram for the truncated normal distributions in this section.

Exercise 7.10: Generate 10,000 samples of size 248 from the normal distribution and discard any negative observations to obtain a truncated normal sample. The size of the truncated normal sample varies. Plot a histogram of the sample sizes. How many samples have fewer than 100 observations?

7.8 References

Bowman, K. O. and Shenton, L. R. (1983), "Johnson's System of Distributions," in S. Kotz, N. L. Johnson, and C. B. Read, eds., *Encyclopedia of Statistical Sciences*, volume 4, 303–314, New York: John Wiley & Sons.

Devroye, L. (1986), *Non-uniform Random Variate Generation*, New York: Springer-Verlag. URL http://luc.devroye.org/rnbookindex.html

Johnson, M. E. (1987), *Multivariate Statistical Simulation*, New York: John Wiley & Sons.

Johnson, N. L. (1949), "Systems of Frequency Curves Generated by Methods of Translation," *Biometrika*, 36, 149–176.

Johnson, N. L., Kotz, S., and Balakrishnan, N. (1994), *Continuous Univariate Distributions*, volume 1, 2nd Edition, New York: John Wiley & Sons.

Robert, C. P. (1995), "Simulation of Truncated Normal Variables," *Statistics and Computing*, 5, 121–125.

Ross, S. M. (2006), *Simulation*, 4th Edition, Orlando, FL: Academic Press.

Tukey, J. W. (1960), "A Survey of Sampling from Contaminated Distributions," in I. Olkin, ed., *Contributions to Probability and Statistics*, Stanford, CA: Stanford University Press.

Chapter 8
Simulating Data from Basic Multivariate Distributions

8.1 Overview of Simulation from Multivariate Distributions

In the classic reference book, *Random Number Generation and Monte Carlo Methods*, the author writes, "Only a few of the standard univariate distributions have standard multivariate extensions" (Gentle 2003, p. 203). Nevertheless, simulating data from multivariate data is an important technique for statistical programmers.

If you have p *uncorrelated* variables, then you can independently simulate each variable by using techniques from the previous chapters. However, the situation becomes more complicated when you need to simulate p variables with a prescribed correlation structure.

This chapter describes how to simulate data from some well-known multivariate distributions, including the multinomial and multivariate normal (MVN) distributions.

8.2 The Multinomial Distribution

There has been considerable research on simulating data from various continuous multivariate distributions. In contrast, there has been less research devoted to simulating data from discrete multivariate distributions. With the exception of the multinomial distribution, some reference books do not mention multivariate discrete distributions at all.

This section describes how to generate random samples from the multinomial distribution. Chapter 9, "Advanced Simulation of Multivariate Data," describes how to simulate correlated binary variates and correlated ordinal variates.

The multinomial distribution is related to the "Table" distribution, which is described in Section 2.4.5. Suppose there are k items in a drawer and p_i is the probability of drawing item i, $i = 1, \ldots, k$, where $\Sigma_i \, p_i = 1$. If you draw N items with replacement, then the number of items chosen for each item makes a single draw from the multinomial distribution. For example, in Section 2.4.5, 100 socks were drawn (with replacement) from a drawer with three colors of socks. A draw might result in 48 black socks, 21 brown socks, and 31 white socks. That triplet of values, $(48, 21, 31)$, is a single observation for the multinomial distribution of three categories (the colors) with the given probabilities.

Notice that if there are only two categories with selection probabilities p and $1 - p$, then the multinomial distribution is equivalent to the binomial distribution with parameter p.

You can sample from the multinomial distribution in SAS/IML software by using the RANDMULTINOMIAL function. The RANDMULTINOMIAL function computes a random sample by using the conditional distribution technique that is described in Section 8.7.1. The syntax of the RANDMULTINOMIAL function is

```
X = RandMultinomial(NumSamples, NumTrials, Prob);
```

where **NumSamples** is the number of observations in the sample, **NumTrials** is the number of independent draws of the k categories, and **Prob** is a $1 \times k$ vector of probabilities that sum to 1.

The following statements generate a 1000×3 matrix, where each row is a random observation from a multinomial distribution. The parameters used here are the same parameters that were used in Section 2.4.5. The first five random draws are shown in Figure 8.1.

```
%let N = 1000;                          /* size of each sample        */
proc iml;
call randseed(4321);                    /* set seed for RandMultinomial */
prob = {0.5 0.2 0.3};
X = RandMultinomial(&N, 100, prob);     /* one sample, N x 3 matrix */

/* print a few results */
c = {"black", "brown", "white"};
first = X[1:5,];
print first[colname=c label="First 5 Obs: Multinomial"];
```

Figure 8.1 Random Draws from Multinomial Distribution

First 5 Obs: Multinomial		
black	brown	white
54	22	24
48	21	31
57	20	23
46	21	33
46	20	34

Notice that the sum across each row is 100 because each row is the frequency distribution that results from drawing 100 socks (with replacement). Although there is variation within each column, elements in the first column are usually close to the expected value of 50, elements in the second column are close to 20, and elements in the third column are close to 30. You can examine the distribution of each component of the multivariate distribution by computing the column means and standard deviations, as shown in Figure 8.2:

```
mean = mean(X);
std = std(X);
corr = corr(X);
print mean[colname=c],
      std[colname=c],
      corr[colname=c rowname=c format=BEST5.];
```

Figure 8.2 Descriptive Statistics of Components of Multinomial Variables

mean		
black	brown	white
50.059	20.003	29.938

std		
black	brown	white
4.7598503	3.9770351	4.481321

corr			
	black	brown	white
black	1	-0.49	-0.63
brown	-0.49	1	-0.37
white	-0.63	-0.37	1

Figure 8.2 shows that the variables are negatively correlated with each other. This makes sense: There are exactly 100 items in each draw, so if you draw more of one color, you will have fewer of the other colors. In fact, the correlation of the component random variables X_i and X_j is $-\sqrt{p_i p_j / ((1 - p_i)(1 - p_j))}$ for $i \neq j$.

You can visualize the multinomial distribution by plotting the first two components against each other, as shown in Figure 8.3. (The third component is always 100 minus the sum of the first two components, so you do not need to plot it.) The multinomial distribution is a discrete distribution whose values are counts, so there is considerable overplotting in a scatter plot of the components. One way to resolve the overplotting is to overlay a kernel density estimate. Areas of high density correspond to areas where there are many overlapping points.

```
/* write multinomial data to SAS data set */
create MN from X[c=c]; append from X; close MN;
quit;

ods graphics on;
proc kde data=MN;
   bivar black brown / plots=ContourScatter;
run;
```

Figure 8.3 Density Plot of Multinomial Components

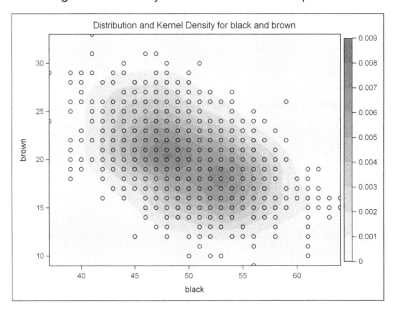

Figure 8.3 shows the negative correlation in the simulated data. It also shows that most of the density is near the population mean $(50, 20)$, although some observations are far from the mean.

Exercise 8.1: A technique that is used to visualize scatter plots that suffer from overplotting is to *jitter* the values by a small amount. Write a DATA step that adds a uniform random value in $[-1/2, 1/2]$ to each component of the multinomial data. Repeat the call to PROC KDE or use PROC SGPLOT to visualize the jittered distribution.

8.3   Multivariate Normal Distributions

Computationally, the multivariate normal distribution is one of the easiest continuous multivariate distributions to simulate. In SAS software, you can simulate MVN data by using the RANDNORMAL function in SAS/IML software or by using PROC SIMNORMAL, which is in SAS/STAT software. In both cases, you need to provide the mean and covariance matrix of the distribution. If, instead, you have a correlation matrix and a set of variances, then you can use the technique in Section 10.2 to create the corresponding covariance matrix.

The MVN distribution is often a "baseline model in Monte Carlo simulations" (Johnson 1987, p. 55) in the sense that many statistical methods perform well on normal data. You can generate results from normal data and compare them with results from nonnormal data.

In the following sections, MVN data are simulated in two ways:

- The RANDNORMAL function in SAS/IML software. Use this technique when you intend to do additional simulation and analysis in PROC IML. For example, in Chapter 11, "Simulating Data for Basic Regression Models," the RANDNORMAL function is used as part of simulation to generate data for a mixed model with correlated random effects. In Section 9.2 and Section 9.3, multivariate normal variates are generated as part of a larger algorithm to generate multivariate binary and ordinal data, respectively.

- The SIMNORMAL function in SAS/STAT software. Use this technique when you need to quickly generate data as input for a prewritten macro or procedure.

You might ask whether it is possible to simulate MVN data by using the DATA step. Yes, it is possible, but it is more complicated than using the SAS/IML language because you have to manually implement matrix multiplication in the DATA step. As shown in Fan et al. (2002, Ch. 4), you can use PROC FACTOR to decompose a correlation matrix into a "factor pattern matrix." You can use this factor pattern matrix to transform uncorrelated random normal variables into multivariate normal variables with the given correlation. Fan et al. (2002) provide a SAS macro to do this. However, calling PROC SIMNORMAL is much easier.

8.3.1   Simulating Multivariate Normal Data in SAS/IML Software

The RANDNORMAL function is available in SAS/IML software; it is not supported by the DATA step. The syntax of the RANDNORMAL function is

```
X = RandNormal(NumSamples, Mean, Cov);
```

where **NumSamples** is the number of observations in the sample, **Mean** is a $1 \times p$ vector, and **Cov** is a $p \times p$ positive definite matrix. (See Section 10.4 for ways to specify a covariance matrix.) The function returns a matrix, **X**, where each row is a random observation from the MVN distribution with the specified mean and covariance.

The following statements simulate 1,000 observations from a trivariate normal distribution. Figure 8.4 displays the first few observations of the simulated data, and the sample mean and covariance.

```
%let N = 1000;                              /* size of each sample */

/* Multivariate normal data */
proc iml;
/* specify the mean and covariance of the population */
Mean = {1, 2, 3};
Cov = {3 2 1,
       2 4 0,
       1 0 5};
call randseed(4321);
X = RandNormal(&N, Mean, Cov);              /* 1000 x 3 matrix     */

/* check the sample mean and sample covariance */
SampleMean = mean(X);                       /* mean of each column */
SampleCov =  cov(X);                        /* sample covariance   */

/* print results */
c = "x1":"x3";
print (X[1:5,])[label="First 5 Obs: MV Normal"];
print SampleMean[colname=c];
print SampleCov[colname=c rowname=c];
```

Figure 8.4 Simulated Data with Sample Mean and Covariance Matrix

First 5 Obs: MV Normal		
3.1489103	2.5584255	1.7839573
2.1945048	2.2662946	-0.113497
-0.172247	-2.353316	4.329263
0.4788255	2.8804713	0.7616427
-0.976523	1.9706425	3.443764

SampleMean		
x1	x2	x3
0.9820782	1.9621798	3.0361992

SampleCov			
	x1	x2	x3
x1	3.0963099	2.0766315	0.9692368
x2	2.0766315	3.8598227	0.0249837
x3	0.9692368	0.0249837	4.7816493

The sample mean is close to the population mean and the sample covariance matrix is close to the population covariance. Just as is true for univariate data, statistics computed on small samples exhibit more variation than statistics on larger samples. Also, the variance of higher-order moments is greater than for lower-order moments.

To visually examine the simulated data, you can write the data to a SAS data set and use the SGSCATTER procedure to create a plot of the univariate and bivariate marginal distributions. Alternatively, you can use the CORR procedure to produce Figure 8.5, as is shown in the following statements. The CORR procedure also produces the sample mean and sample covariance, but these tables are not shown.

```
/* write SAS/IML matrix to SAS data set for plotting */
create MVN from X[colname=c];  append from X;  close MVN;
quit;

/* create scatter plot matrix of simulated data */
ods graphics on;
proc corr data=MVN COV plots(maxpoints=NONE)=matrix(histogram);
   var x:;
run;
```

Figure 8.5 Univariate and Bivariate Marginal Distributions for Simulated Multivariate Normal Data

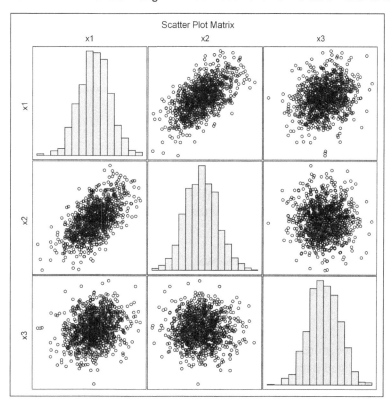

Figure 8.5 shows that the marginal distribution for each variable (displayed as histograms on the diagonal) appears to be normal, as do the pairwise bivariate distributions (displayed as scatter plots). This is characteristic of MVN data: All marginal distributions are normally distributed.

In addition to the graphical checks, there are rigorous goodness-of-fit tests to determine whether data are likely to have come from a multivariate normal distribution. You can search the support.sas.com Web site to find the %MULTNORM macro, which performs tests for multivariate normality. The %MULTNORM macro also computes and plots the squared Mahalanobis distances of the observations to the mean vector. For p-dimensional MVN data, the squared distances are distributed as chi-square with p degrees of freedom. Consequently, a plot of the squared distance versus quantiles of a chi-square distribution will fall along a straight line for data that are multivariate normal.

Exercise 8.2: The numeric variables in the Sashelp.Iris data set are SepalLength, SepalWidth, PetalLength, and PetalWidth. There are three species of flowers in the data. Use PROC CORR to visualize these variables for the "Virginica" species. Do the data appear to be multivariate normal? Repeat the analysis for the "Setosa" species.

8.3.2   Simulating Multivariate Normal Data in SAS/STAT Software

A second way to generate MVN data is to use the SIMNORMAL procedure in SAS/STAT software. As is shown in Section 8.3.1, you must provide the mean and covariance of the distribution in order to simulate the data. For PROC SIMNORMAL, this is accomplished by creating a TYPE=COV data set, as shown in the following example:

```
/* create a TYPE=COV data set */
data MyCov(type=COV);
input _TYPE_ $ 1-8 _NAME_ $ 9-16 x1 x2 x3;
datalines;
COV      x1      3 2 1
COV      x2      2 4 0
COV      x3      1 0 5
MEAN             1 2 3
run;
```

The data set specifies the same mean vector and covariance matrix as the example in Section 8.3.1. This data set is used as the input data for the SIMNORMAL procedure. For example, the following call generates 1,000 random observations from the MVN distribution with the specified mean and covariance:

```
proc simnormal data=MyCov outsim=MVN
               nr = 1000            /* size of sample     */
               seed = 12345;        /* random number seed */
    var x1-x3;
run;
```

The SIMNORMAL procedure does not produce any tables; it only creates the output data set that contains the simulated data. You can run PROC CORR and the %MULTNORM macro to convince yourself that the output of PROC SIMNORMAL is multivariate normal from the specified distribution.

The SIMNORMAL procedure also enables you to perform conditional simulation from an MVN distribution. Given an MVN distribution with p variables, you can specify the values of $k < p$ variables and simulate the remaining $p - k$, conditioned on the specified values. See Section 8.6 for more information about conditional simulation.

Exercise 8.3: Use PROC CORR and the %MULTNORM macro to analyze the simulated data in the MVN data set.

8.4   Generating Data from Other Multivariate Distributions

SAS/IML software supports several other multivariate distributions:

- The RANDDIRICHLET function generates a random sample from a Dirichlet distribution, which is not used in this book.

- The RANDMVT function generates a random sample from a multivariate Student's t distribution (Kotz and Nadarajah 2004).

- The RANDWISHART function, which is described in Section 10.5, generates a random sample from a Wishart distribution.

Johnson (1987) presents many other (lesser-known) multivariate distributions and describes simulation techniques for generating multivariate data. There are also multivariate distributions mentioned in Gentle (2003).

The multivariate Student's t distribution is useful when you want to simulate data from a multivariate distribution that is similar to the MVN distribution, but has fatter tails. The multivariate t distribution with ν degrees of freedom has marginal distributions that are univariate t with ν degrees of freedom.

The SAS/IML RANDMVT function simulates data from a multivariate t distribution. The syntax of the RANDMVT function is

```
X = RandMVT(NumSamples, DF, Mean, S);
```

where **NumSamples** is the number of observations in the sample, **DF** is the degrees of freedom, **Mean** is a $1 \times p$ vector, and **S** is a $p \times p$ positive definite matrix. Given the matrix S, the population covariance is $\frac{\nu}{\nu-2}S$ where $\nu > 2$ is the degrees of freedom. For large values of ν, the multivariate t distribution is approximately multivariate normal.

The following statements generate a 100×3 matrix, where each row is a random observation from a multivariate t distribution with 4 degrees of freedom:

```
proc iml;
/* specify population mean and covariance */
Mean = {1, 2, 3};
Cov = {3 2 1,
       2 4 0,
       1 0 5};
call randseed(4321);
X = RandMVT(100, 4, Mean, Cov);   /* 100 draws; 4 degrees of freedom */
```

Exercise 8.4: Use PROC CORR to compute the sample correlations and to visualize the distribution of the multivariate t data with four degrees of freedom. Can you see evidence that the tails of the t distribution are heavier than for the normal distribution?

8.5   Mixtures of Multivariate Distributions

As was described in Section 7.5 for univariate distributions, you can simulate multivariate data from a mixture of other distributions. This is useful, for example, for simulating clustered data and data for discriminant analysis.

8.5.1   The Multivariate Contaminated Normal Distribution

The multivariate contaminated normal distribution is similar to the univariate version that is presented in Section 7.5.2. Given a mean vector μ and a covariance matrix Σ, you can sample from an MVN(μ, Σ) distribution with probability $1 - p$ and from an MVN$(\mu, k^2\Sigma)$ distribution with probability p, where $k > 1$ is a constant that represents the size of the contaminated component. This results in a distribution with heavier tails than normality. This is also a convenient way to generate data with outliers.

The following SAS/IML statements call the RANDNORMAL function to generate N_1 observations from MVN(μ, Σ) and $N - N_1$ observations from MVN$(\mu, k^2\Sigma)$, where N_1 is chosen randomly from a binomial distribution with parameter $1 - p$:

```
/* create multivariate contaminated normal distrib */
%let N = 100;
proc iml;
mu =   {0 0 0};                          /* vector of means        */
Cov = {10   3   -2,
        3   6    1,
       -2   1    2};
k2 = 100;                                /* contamination factor   */
p = 0.1;                                 /* prob of contamination  */

/* generate contaminated normal (mixture) distribution */
call randseed(1);
call randgen(N1, "Binomial", 1-p, &N); /* N1 unallocated ==> scalar */

X = j(&N, ncol(mu));
X[1:N1,] = RandNormal(N1, mu, Cov);                /* uncontaminated */
X[N1+1:&N,] = RandNormal(&N-N1, mu, k2*Cov);       /* contaminated   */
```

Notice that the program only makes two calls to the RANDNORMAL function. This is more efficient than writing a DO loop that iteratively draws a Bernoulli random variable and simulates a single observation from the selected distribution.

You can create a SAS data set and use PROC CORR to visualize the simulated data. Figure 8.7 shows the scatter plot matrix of the three variables.

```
/* write SAS data set */
create Contam from X[c=('x1':'x3')];   append from X;   close Contam;
quit;

proc corr data=Contam cov plots=matrix(histogram);
   var x1-x3;
run;
```

Figure 8.6 Analysis of Contaminated Multivariate Normal Data

The CORR Procedure

Covariance Matrix, DF = 99			
	x1	**x2**	**x3**
x1	268.2518872	51.4941731	-75.2288082
x2	51.4941731	135.6709963	29.2113495
x3	-75.2288082	29.2113495	44.3598676

Simple Statistics						
Variable	**N**	**Mean**	**Std Dev**	**Sum**	**Minimum**	**Maximum**
x1	100	2.15309	16.37840	215.30913	-70.20609	75.03301
x2	100	-0.53016	11.64779	-53.01628	-51.41821	54.05912
x3	100	-1.13626	6.66032	-113.62624	-43.98791	13.69675

Figure 8.7 Multivariate Normal Data with 10% Contamination

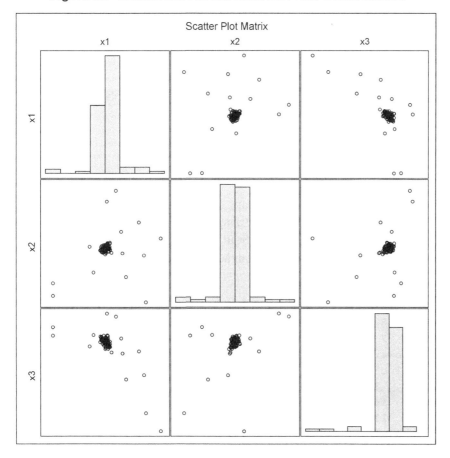

The plots show the high density of observations near the center of each scatter plot. These observations were primarily drawn from the uncontaminated distribution. The remaining observations are from the distribution with the larger variance.

Notice that the covariance estimates that are found by the CORR procedure (see Figure 8.6) are far from the values of the uncontaminated parameters. You can use the MCD subroutine in the SAS/IML language or the ROBUSTREG routine in SAS/STAT software to estimate the covariance parameters in ways that are robust to the presence of outliers.

Exercise 8.5: Use the %MULTNORM macro to test the data in the Contam data set for multivariate normality. Does the chi-square Q-Q plot reveal the presence of two distributions?

8.5.2   Mixtures of Multivariate Normal Distributions

It is easy to extend the contaminated normal example to simulate data from a mixture of k MVN distributions. Suppose that you want to generate N observations from a mixture distribution where each observation has probability π_i of being drawn from $\text{MVN}(\mu_i, \Sigma_i), i = 1 \ldots k$. The following program uses the multinomial distribution (see Section 8.2) to randomly produce N_i, which is the number of observations to be simulated from the ith component distribution:

```
proc iml;
call randseed(12345);
pi = {0.35 0.5 0.15};              /* mixing probs for k groups */
NumObs = 100;                      /* total num obs to sample   */
N = RandMultinomial(1, NumObs, pi);
print N;
```

Figure 8.8 A Draw from the Multinomial Distribution

N		
26	51	23

The following statements continue the program. The mean vectors and covariance matrices are stored as rows of a matrix, and only the lower triangular portion of each covariance matrix is stored. A row is converted to a matrix by calling the SQRVECH function; see Section 10.4.2. For the ith group, you can generate `N[i]` observations from $\text{MVN}(\mu_i, \Sigma_i), i = 1 \ldots 3$.

```
varNames={"x1" "x2" "x3"};
mu =    {32   16    5,              /* means of Group 1      */
          30    8    4,             /* means of Group 2      */
          49    7    5};            /* means of Group 3      */
/* specify lower-triangular within-group covariances */
/*    c11 c21 c31 c22 c32 c33 */
Cov = {17  7    3   5   1   1,      /* cov of Group 1        */
        90 27  16   9   5   4,      /* cov of Group 2        */
       103 16  11   4   2   2};     /* cov of Group 3        */
```

```
/* generate mixture distribution: Sample from
   MVN(mu[i,], Cov[i,]) with probability pi[i] */
p = ncol(pi);                                    /* number of variables  */
X = j(NumObs, p);
Group = j(NumObs, 1);
b = 1;                                           /* beginning index      */
do i = 1 to p;
   e = b + N[i] - 1;                             /* ending index         */
   c = sqrvech(Cov[i,]);                         /* cov of group (dense) */
   X[b:e, ] = RandNormal(N[i], mu[i,], c);       /* i_th MVN sample      */
   Group[b:e] = i;
   b = e + 1;                                    /* next group starts at this index */
end;

/* save to data set */
Y = Group || X;
create F from Y[c=("Group" || varNames)];  append from Y;  close F;
quit;

proc sgscatter data=F;
   compare y=x2 x=(x1 x3) / group=Group markerattrs=(Size=12);
run;
```

Figure 8.9 Grouped Data from MVN Distributions

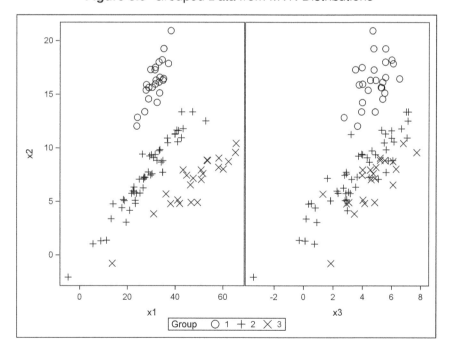

The SGSCATTER procedure is used to visualize the simulated data. The graphs show that there is separation between the groups in the x_1 and x_2 variables because the first two components of the mean vectors are far apart relative to the covariances. However, there is little separation in the x_2 and x_3 variables.

Exercise 8.6: Encapsulate the SAS/IML statements into a function that generates N random observations from a mixture of MVN distributions.

8.6   Conditional Multivariate Normal Distributions

A *conditional distribution* is the distribution that results when several variables are fixed at some specific value. If $f(x_1, x_2, \ldots, x_p)$ is a probability distribution for p variables, then the conditional probability distribution at $x_p = a$ is $f(x_1, x_2, \ldots, a)$. That is, you substitute the known value. Geometrically, the conditional distribution is obtained by slicing the full distribution with the hyperplane $x_p = a$. You can also form a conditional distribution by specifying the value of more than one variable. For example, you can specify a k-dimensional conditional distribution by $g(x_1, x_2, \ldots, x_k) = f(x_1, x_2, \ldots, k, a_{k+1}, \ldots, a_p)$.

For some multivariate distributions, it is difficult to sample from a conditional distribution. Although it is easy to obtain the density function (just substitute the specified values), the resulting density might not be a familiar "named" distribution such as the normal, gamma, or beta distribution.

However, conditional distributions for the MVN distribution are easy to generate because every conditional distribution is also MVN and you can compute the conditional mean and covariance matrix directly (Johnson 1987, p. 50). In particular, let $X \sim \text{MVN}(\mu, \Sigma)$, where

$$X = \begin{bmatrix} X_1 \\ X_2 \end{bmatrix}, \quad \mu = \begin{bmatrix} \mu_1 \\ \mu_2 \end{bmatrix}, \quad \Sigma = \begin{pmatrix} \Sigma_{11} & \Sigma_{12} \\ \Sigma_{21} & \Sigma_{22} \end{pmatrix}$$

The vector X_1 is $k \times 1$, X_2 is $(p-k) \times 1$, Σ_{11} is $k \times k$, Σ_{22} is $(p-k) \times (p-k)$, and $\Sigma_{12} = \Sigma_{21}'$ is $k \times (p-k)$.

The distribution of X_1, given that $X_2 = v$, is $N_k(\tilde{\mu}, \tilde{\Sigma})$, where

$$\tilde{\mu} = \mu_1 + \Sigma_{12} \Sigma_{22}^{-1} (v - \mu_2)$$

and

$$\tilde{\Sigma} = \Sigma_{11} - \Sigma_{12} \Sigma_{22}^{-1} \Sigma_{21}$$

You can use the SIMNORMAL procedure to carry out conditional simulation. See the documentation in the *SAS/STAT User's Guide* for details and an example. In addition, the following SAS/IML program defines a module to compute the conditional mean and covariance matrix, given values for the last k variables, $0 < k < p$.

```
proc iml;
/* Given a p-dimensional MVN distribution and p-k fixed values for
   the variables x_{k+1},...,x_p, return the conditional mean and
   covariance for first k variables, conditioned on the last p-k
   variables. The conditional mean is returned as a column vector. */
start CondMVNMeanCov(m, S, _mu, Sigma, _v);
   mu = colvec(_mu);   v = colvec(_v);
   p = nrow(mu);       k = p - nrow(v);
```

```
   mu1 = mu[1:k];
   mu2 = mu[k+1:p];
   Sigma11 = Sigma[1:k,  1:k];
   Sigma12 = Sigma[1:k,  k+1:p]; *Sigma21 = T(Sigma12);
   Sigma22 = Sigma[k+1:p,  k+1:p];
   m = mu1 + Sigma12*solve(Sigma22, (v - mu2));
   S = Sigma11 - Sigma12*solve(Sigma22, Sigma12`);
finish;
```

The module returns the conditional mean in the first argument and the conditional covariance in the second argument. To demonstrate how the module works, the following example sets $x_3 = 2$. The conditional mean and covariance are shown in Figure 8.10.

```
mu = {1 2 3};                            /* 3D MVN example */
Sigma = {3 2 1,
         2 4 0,
         1 0 9};
v3 = 2;                                  /* value of x3    */
run CondMVNMeanCov(m, c, mu, Sigma, v3);
print m c;
```

Figure 8.10 Conditional Mean and Covariance

m	c	
0.8888889	2.8888889	2
2	2	4

Notice that the second component of the conditional mean is 2, which is the same as for the unconditional mean because X_2 and X_3 are uncorrelated. Similarly, the second column of the conditional covariance is the same ({2 4}) as for the unconditional covariance. However, the first component of the conditional mean is different from the unconditional value because X_1 and X_3 are correlated.

You can use the conditional means and covariances to simulate observations from the conditional distribution: Simply call the RANDNORMAL function with the conditional parameter values, as shown in the following example:

```
/* Given a p-dimensional MVN distribution and p-k fixed values
   for the variables x_{k+1},...,x_p, simulate first k
   variables conditioned on the last p-k variables. */
start CondMVN(N, mu, Sigma, v);
   run CondMVNMeanCov(m, S, mu, Sigma, v);
   return( RandNormal(N, m`, S) );          /* m` is row vector */
finish;

call randseed(1234);
N = 1000;
z = CondMVN(N, mu, Sigma, v3);   /* simulate 2D conditional distrib */
```

You can write the data to a SAS data set and plot it by using the SGPLOT procedure. The following statements also overlay bivariate normal probability ellipses on the simulated data:

```
varNames = "x1":"x2";
create mvn2 from z[c=varNames]; append from z; close mvn2;
quit;

proc sgplot data=mvn2 noautolegend;
   scatter x=x1 y=x2;
   ellipse x=x1 y=x2 / alpha=0.05;
   ellipse x=x1 y=x2 / alpha=0.1;
   ellipse x=x1 y=x2 / alpha=0.2;
   ellipse x=x1 y=x2 / alpha=0.5;
run;
```

Figure 8.11 Simulated Data from Conditional MVN Distribution

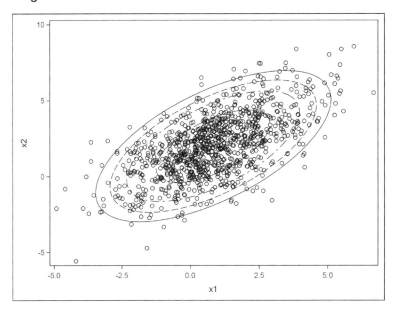

Exercise 8.7: Call the CondMVNMeanCov function with the input `v = {2, 3}`. What is the mean and variance of the resulting conditional distribution? Call the CondMVN function with the same input. Simulate 1,000 observations from the conditional distribution and create a histogram of the simulated data.

8.7 Methods for Generating Data from Multivariate Distributions

Generating data from multivariate distributions with correlated components and specified marginal distributions is hard. There have been many papers written about how to simulate from multivariate distributions that have certain properties. Relatively few general techniques exist for simulating data that have both a target correlation structure and specified marginal distributions. One reason is that if you specify the marginal distributions arbitrarily, then some correlation structures are unobtainable.

The simulation literature contains three general techniques that are used frequently. They are the conditional distribution technique, the transformation technique, and the copula technique, which combines features of the first two techniques. The conditional distribution and transformation techniques are described in Johnson (1987), which is used as the basis for the notation and ideas in this section. The copula technique is described in Section 9.5.

8.7.1 The Conditional Distribution Technique

The goal of a multivariate simulation is to generate observations from a p-dimensional distribution. Each observation is the realization of some random vector $X = (X_1, X_2, \ldots, X_p)$. The conditional distribution technique converts the problem into a sequence of p one-dimensional problems. You generate an observation $x = (x_1, x_2, \ldots, x_p)$ by doing the following:

1. Generate x_1 from the marginal distribution of X_1.

2. For $k = 2, \ldots, p$, generate x_k from the conditional distribution of X_k, given $X_1 = x_1, X_2 = x_2, \ldots, X_{k-1} = x_{k-1}$.

One problem with this method is that it might be difficult to compute a random variate from the conditional distributions. You might not be able to express the conditional distributions in an exact form because they can be complicated functions of the previous variates. Another problem is that conditional techniques tend to be much slower than direct techniques. However, an advantage is that the conditional technique is broadly applicable, even in very high dimensions. The conditional distribution technique is the basis for Gibbs sampling, which is available for Bayesian modeling in the MCMC procedure in SAS/STAT software.

As an example, you can use the formulas shown in Section 8.6 to generate data from a conditional MVN distribution by using the CondMVNMeanCov module multiple times. For example, for three-dimensional data, you can generate a single observation by doing the following:

1. Generate x_3 from a univariate normal distribution with mean μ_3 and variance Σ_{33}. Pass this x_3 value into the CondMVNMeanCov module to get a two-dimensional conditional mean, ν, and covariance matrix, A.

2. Generate x_2 from a univariate normal distribution with mean ν_2 and variance A_{22}. Pass this x_2 value into the CondMVNMeanCov module to get a one-dimensional conditional mean and covariance matrix.

3. Generate x_1 from a univariate normal distribution with the mean and variance found in the previous step.

For the MVN distribution, this technique is not as efficient as the transformation technique that is shown in Section 8.7.2. However, it demonstrates that the conditional distribution technique enables you to sample from a multivariate distribution by sampling one variable at a time.

Exercise 8.8: Write a SAS/IML function that uses the conditional technique to simulate from the three-dimensional MVN distribution that is specified in Section 8.6. Compare the time it takes to simulate 10,000 observations by using the conditional distribution technique with the time it takes to generate the same number of observations by using the RANDNORMAL function.

8.7.2   The Transformation Technique

Often the conditional distributions of a multivariate distribution are difficult to compute. A second technique for generating multivariate samples is to generate p values from independent univariate distributions and transform those values to have the properties of the desired multivariate distribution. This *transformation technique* is widely used. For example, many of the algorithms in Devroye (1986) use this technique.

Section 7.3 provides univariate examples that use the transformation technique. The beauty of the technique is that many distributions can be generated by using independent uniform or normal distribution as the basis for the (pre-transformed) data. For example, Johnson (1987) mentions that the multivariate Cauchy distribution can also be generated from standard normal distributions and a Gamma(1/2) distribution. Specifically, if Z_i is a standard normal variable, $i = 1, \ldots, p$, and Y is a Gamma(1/2) variable, then you can generate the multivariate Cauchy distribution from $Z_i / \sqrt{2Y}$ as shown in the following SAS/IML module. Notice that the module is vectorized: All N observations for all p normal variables are generated in a single call, as are all N Gamma(1/2) variates.

```
proc iml;
/* Sample from a multivariate Cauchy distribution */
start RandMVCauchy(N, p);
   z = j(N,p,0);  y = j(N,1);          /* allocate matrix and vector */
   call randgen(z, "Normal");
   call randgen(y, "Gamma", 0.5);      /* alpha=0.5, unit scale      */
   return( z / sqrt(2*y) );
finish;

/* call the function to generate multivariate Cauchy variates */
N=1000; p = 3;
x = RandMVCauchy(N, p);
```

The transformation technique generates the data directly; there is no iteration as in the conditional simulation technique.

8.8   The Cholesky Transformation

The RANDNORMAL function that is discussed in Section 8.3.1 uses a technique known as the *Cholesky transformation* to simulate MVN data from a population with a given covariance structure. This section describes the Cholesky transformation and its geometric properties. This transformation is often used as part of an algorithm to generate correlated multivariate data from a wide range of distributions.

A covariance matrix, Σ, contains the covariances between random variables: $\Sigma_{ij} = \text{Cov}(X_i, X_j)$. You can transform a set of uncorrelated variables into variables with the given covariances. The transformation that accomplishes this task is called the Cholesky transformation; it is represented by a matrix that is a "square root" of the covariance matrix.

Any covariance matrix, Σ, can be factored uniquely into a product $\Sigma = U^T U$, where U is an upper triangular matrix with positive diagonal entries and the superscript denotes matrix transpose. The matrix U is the Cholesky (or "square root") matrix. Some researchers such as Golub and Van Loan (Golub and Van Loan 1989, p. 143) prefer to work with lower triangular matrices. If you define

$L = U^T$, then $\Sigma = LL^T$. Golub and Van Loan provide a proof of the Cholesky decomposition, as well as various ways to compute it. In SAS/IML software, the Cholesky decomposition is computed by using the ROOT function.

The Cholesky matrix transforms uncorrelated variables into variables whose variances and covariances are given by Σ. In particular, if you generate p standard normal variates, the Cholesky transformation maps the variables into variables for the MVN distribution with covariance matrix Σ and centered at the origin (denoted MVN(0, Σ)).

For a simple example, suppose that you want to generate MVN data that are uncorrelated, but have non-unit variance. The covariance matrix for this situation is the diagonal matrix of variances: $\Sigma = \text{diag}(\sigma_1^2, \ldots, \sigma_p^2)$. The square root of Σ is the diagonal matrix D that consists of the standard deviations: $\Sigma = D^T D$ where $D = \text{diag}(\sigma_1, \ldots, \sigma_p)$.

Geometrically, the D matrix scales each coordinate direction independently of the other directions. This is shown in Figure 8.12. The horizontal axis is scaled by a factor of 3, whereas the vertical axis is unchanged (scale factor of 1). The transformation is $D = \text{diag}(3, 1)$, which corresponds to a covariance matrix of diag(9, 1). If you think of the circles in the top image as being probability contours for the multivariate distribution MVN(0, I), then the bottom shows the corresponding probability ellipses for the distribution MVN(0, D).

Figure 8.12 The Geometry of a Diagonal Transformation

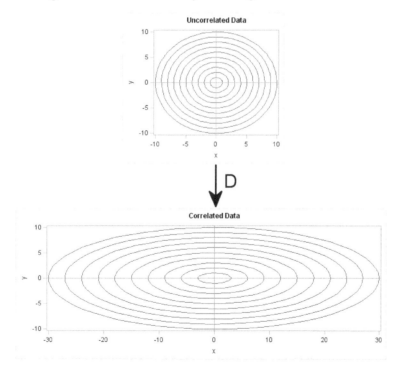

In the general case, a covariance matrix contains off-diagonal elements. The geometry of the Cholesky transformation is similar to the pure-scaling case shown previously, but also shears and rotates the top image.

Computing a Cholesky matrix for a general covariance matrix is not as simple as computing a Cholesky matrix for a diagonal covariance matrix. In SAS/IML software, the ROOT function returns a matrix U such that the product $U^T U$ equals the covariance matrix, and U is an upper triangular matrix with positive diagonal entries. The following statements compute a Cholesky matrix in PROC IML:

```
proc iml;
Sigma = {9 1,
         1 1};
U = root(Sigma);
print U[format=BEST5.];                      /* U`*U = Sigma */
```

Figure 8.13 A Cholesky Matrix

U	
3	0.333
0	0.943

You can use the Cholesky matrix to create correlations among random variables. For example, suppose that X and Y are independent standard normal variables. The matrix U (or its transpose, $L = U^T$) can be used to create new variables Z and W such that the covariance of Z and W equals Σ. The following SAS/IML statements generate X and Y as rows of the matrix **xy**. That is, each column is a point (x, y). (Usually the columns of a matrix are used to store variables, but transposing **xy** makes the linear algebra simpler.) The statements then map each (x, y) pair to a new point, (z, w), and compute the sample covariance of the Z and W variables. As promised, the sample covariance is close to Σ, which is the covariance of the underlying population.

```
/* generate x,y ~ N(0,1), corr(x,y)=0 */
call randseed(12345);
xy = j(2, 1000);
call randgen(xy, "Normal");                  /* each col is indep N(0,1) */

L = U`;
zw = L * xy;              /* Cholesky transformation induces correlation */
cov = cov(zw`);          /* check covariance of transformed variables   */
print cov[format=BEST5.];
```

Figure 8.14 Sample Covariance Matrix

cov	
9.169	1.046
1.046	0.956

Figure 8.15 shows the geometry of the transformation in terms of the data and in terms of probability ellipses. The top image is a scatter plot of the X and Y variables. Notice that they are uncorrelated and that the probability ellipses are circles. The bottom image is a scatter plot of the Z and W variables. Notice that they are correlated and the probability contours are ellipses that are tilted with respect to the coordinate axes. The bottom image is the transformation under L of the points and circles in the top image.

Figure 8.15 The Geometry of a Cholesky Transformation

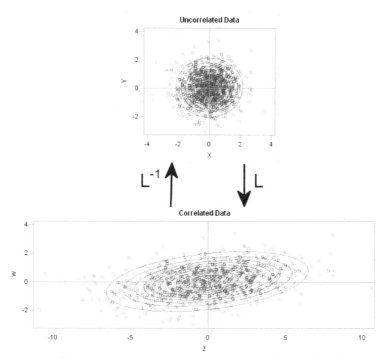

It is also possible to "uncorrelate" correlated variables, and the transformation that uncorrelates variables is the inverse of the Cholesky transformation. Specifically, if you generate data from MVN(0,Σ), you can uncorrelate the data by applying L^{-1}. In SAS/IML software, you might be tempted to use the INV function to compute an explicit matrix inverse, but this is not efficient (Wicklin 2010, p. 372). Because L is a lower triangular matrix, it is more efficient to use the TRISOLV function, as shown in the following statements:

```
/* Start with MVN(0, Sigma) data. Apply inverse of L. */
zw = T( RandNormal(1000, {0, 0}, Sigma) );
xy = trisolv(4, L, zw);            /* more efficient than solve(L,zw) */

/* Did we succeed in uncorrelating the data? Compute covariance. */
tcov = cov(xy`);
print tcov[format=5.3];
```

Figure 8.16 Covariance of Transformed Data

tcov	
1.059	0.006
0.006	1.039

Notice that Figure 8.16 shows that the covariance matrix of the transformed data is close to the identity matrix. The inverse Cholesky transformation has "uncorrelated" the data.

The TRISOLV function, which uses back-substitution to solve the linear system, is extremely fast. Anytime you are trying to solve a linear system that involves a covariance matrix, you should compute the Cholesky factor and use back-substitution to solve the system.

8.9 The Spectral Decomposition

The Cholesky factor is just one way to obtain a "square root" matrix of a correlation matrix. Another option is the *spectral decomposition*, which is also known as the *eigenvalue decomposition*. Given a correlation matrix, R, you can factor R as $R = UDU'$ where U is the matrix of eigenvectors and D is a diagonal matrix that contains the eigenvalues. Then the matrix $H = D^{1/2}U'$ (called the *factor pattern matrix*) is a square root matrix of R because $H'H = R$. Sometimes it is more convenient to work with transposes such as $F = H' = UD^{1/2}$.

A factor pattern matrix can be computed by PROC FACTOR or in the SAS/IML language as part of an eigenvalue decomposition of the correlation matrix. For example, the following example is modified from Fan et al. (2002, p. 72):

```
data A(type=corr);
_type_='CORR';
input x1-x3;
cards;
1.0  .   .
0.7 1.0  .
0.2 0.4 1.0
;
run;

/* obtain factor pattern matrix from PROC FACTOR */
proc factor data=A N=3 eigenvectors;
   ods select FactorPattern;
run;

/* Perform the same computation in SAS/IML language */
proc iml;
R = {1.0 0.7 0.2,
     0.7 1.0 0.4,
     0.2 0.4 1.0};

/* factor pattern matrix via the eigenvalue decomp.
   R = U*diag(D)*U` = H`*H = F*F` */
call eigen(D, U, R);
F = sqrt(D`) # U;                    /* F is returned by PROC FACTOR */
Verify = F*F`;
print F[format=8.5] Verify;
```

Figure 8.17 Factor Pattern Matrix

The FACTOR Procedure
Initial Factor Method: Principal Components

Factor Pattern			
	Factor1	Factor2	Factor3
x1	0.84267	-0.42498	0.33062
x2	0.91671	-0.12476	-0.37958
x3	0.59317	0.79654	0.11694

F			Verify		
0.84267	-0.42498	0.33062	1	0.7	0.2
0.91671	-0.12476	-0.37958	0.7	1	0.4
0.59317	0.79654	0.11694	0.2	0.4	1

The factor pattern matrix H is a square-root matrix for R. Consequently, H can be used to transform uncorrelated standard normal variates into correlated multivariate normal variates. For example, the following SAS/IML statements simulate from an MVN distribution where the variables are correlated according to the R matrix. Figure 8.18 shows that the columns of **x** are correlated and the sample correlation is close to **R**. The spectral decomposition is used in Section 10.8 and Section 16.11.

```
z = j(1000, 3);
call randgen(z, "Normal");    /* uncorrelated normal obs: z~MVN(0,I) */

/* Compute x`=F*z` or its transpose x=z*F` */
x = z*F`;                      /* x~MVN(0,R) where R=FF`= corr matrix */
corr = corr(x);                /* sample correlation is close to R    */
print corr[format=5.3];
```

Figure 8.18 Sample Correlation Matrix of Simulated Data

corr		
1.000	0.719	0.208
0.719	1.000	0.407
0.208	0.407	1.000

8.10   References

Devroye, L. (1986), *Non-uniform Random Variate Generation*, New York: Springer-Verlag.
URL http://luc.devroye.org/rnbookindex.html

Fan, X., Felsovályi, A., Sivo, S. A., and Keenan, S. C. (2002), *SAS for Monte Carlo Studies: A Guide for Quantitative Researchers*, Cary, NC: SAS Institute Inc.

Gentle, J. E. (2003), *Random Number Generation and Monte Carlo Methods*, 2nd Edition, Berlin: Springer-Verlag.

Golub, G. H. and Van Loan, C. F. (1989), *Matrix Computations*, 2nd Edition, Baltimore: Johns Hopkins University Press.

Johnson, M. E. (1987), *Multivariate Statistical Simulation*, New York: John Wiley & Sons.

Kotz, S. and Nadarajah, S. (2004), *Multivariate t Distributions and Their Applications*, Cambridge: Cambridge University Press.

Wicklin, R. (2010), *Statistical Programming with SAS/IML Software*, Cary, NC: SAS Institute Inc.

Chapter 9
Advanced Simulation of Multivariate Data

Contents

9.1 Overview of Advanced Multivariate Simulation

The previous chapter describes how to simulate data from commonly encountered multivariate distributions. This chapter describes how to simulate correlated data for more specialized distributions.

An interesting problem in multivariate simulation is the following. Suppose that you want to specify the marginal distribution of each of p variables *and* you also want to specify the correlations between variables. For example, you might want the first random variable to follow a gamma distribution, the second variable to be exponential, and the third to be normally distributed. Furthermore, you want to specify the correlations between the variables, such as $\rho_{12} = 0.1$, $\rho_{13} = -0.2$, and $\rho_{23} = 0.3$.

In general, this is a difficult problem. To further complicate matters, there are some combinations of marginal distributions and correlations that cannot be mathematically satisfied. That is, when the marginal distributions hold, some correlations are impossible. However, if the specified correlations are feasible, then the *copula* technique might enable you to sample from a multivariate distribution with the given conditions. Copula techniques are heavily used in the fields of finance and econometrics.

Although it is common practice to separately model the marginal distribution and the correlation structure, you should think carefully about whether your model makes mathematical sense. As Schabenberger and Gotway (2005) write: "It may be tempting to combine models for mean and covariance structure that maybe should not be considered in the same breath. The resulting model may be vacuous." In other words, a joint distribution function with the specified marginal distributions and covariance structure might not exist. Schabenberger and Gotway opine that it is "sloppy" to ask whether you can simulate behavior for which "no mechanism comes to mind that could generate this behavior." Those who proceed in spite of these cautions might seek justification in George Box's famous quote, "All models are wrong, but some are useful."

This chapter begins with a detailed description of how to simulate from two discrete distributions: the multivariate binary and multivariate ordinal distributions. You do not need to learn all of the details in order to use the SAS programs that simulate from these distributions. However, the literature is full of similar constructions, and the author hopes that a careful step-by-step description will help you understand how to simulate data from multivariate distributions that are not included in this book.

9.2   Generating Multivariate Binary Variates

Emrich and Piedmonte (1991) describe a straightforward algorithm for generating d binary variables, $X_1, X_2, \ldots, X_d$, such that

- The probability of success, p_i, is specified for each Bernoulli variable, $X_i, i = 1, \ldots, d$. The corresponding probabilities of failure are $q_i = 1 - p_i$.

- The Pearson correlation, $\text{corr}(X_j, X_k) = \delta_{jk}$, is specified for each $j = 1, \ldots, d$ and $k = i + 1, \ldots, d$. For convenience, let Δ be the specified correlation matrix.

As noted by Emrich and Piedmonte (1991), the correlation $\text{corr}(X_j, X_k)$ for two binary variables is constrained by the expected values of X_j and X_k. Specifically, given p_j and p_k, the feasible correlation is in the range $[L, U]$, where $L = \max(-\sqrt{p_j p_k/q_j q_k}, -\sqrt{q_j q_k/p_j p_k})$, $U = \min(-\sqrt{p_j q_k/p_k q_j}, -\sqrt{p_k q_j/p_j q_k})$.

The Emrich-Piedmonte algorithm consists of the following steps:

1. (Optional) Check the specified parameters to ensure that it is feasible to solve the problem.

2. For each j and k, solve the equation $\Phi\left(z(p_j), z(p_k), \rho_{jk}\right) - \delta_{jk}\sqrt{p_j q_j p_k q_k} - p_j p_k = 0$ for ρ_{jk}, where Φ is the cumulative distribution function for the standard bivariate normal distribution and where $z(\cdot)$ is the quantile function for the univariate standard normal distribution. That is, given the desired correlation matrix, Δ, find an "intermediate" correlation matrix, R, with off-diagonal elements equal to ρ_{jk}.

3. Generate multivariate normal variates: $(X_1, \ldots, X_k)$ with mean 0 and correlation matrix R.

4. Generate binary variates: $B_j = 1$ when $X_j < z(p_j)$; otherwise, $B_j = 0$.

The algorithm requires solving an equation in Step 2 that involves the bivariate normal cumulative distribution. You have to solve the equation $d(d-1)/2$ times, which is computationally expensive, but you can use the solution to generate arbitrarily many binary variates. The algorithm has the advantage that it can handle arbitrary correlation structures. Other researchers (see Oman (2009) and references therein) have proposed faster algorithms to generate multivariate binary variables when the correlation structure is a particular special form, such as AR(1) correlation.

Because the equation in Step 2 is solved pairwise for each ρ_{jk}, the matrix R might not be positive definite.

9.2.1   Checking for Feasible Parameters

Step 1 of the algorithm is to check that the pairwise correlations and the marginal probabilities are consistent with each other. Given a pair of marginal probabilities, p_i and p_j, not every pairwise correlation in $[-1, 1]$ can be achieved. The following SAS/IML function verifies that the correlations for the Emrich-Piedmonte algorithm are feasible:

```
proc iml;
/* Let X1, X2,...,Xd be binary variables, let
   p = (p1,p2,...,pd) the their expected values and let
   Delta be the d x d matrix of correlations.
   This function returns 1 if p and Delta are feasible for binary
   variables. The function also computes lower and upper bounds on the
   correlations, and returns them in LBound and UBound, respectively */
start CheckMVBinaryParams(LBound, UBound, _p, Delta);
   p = rowvec(_p);    q = 1 - p;         /* make p a row vector     */
   d = ncol(p);                          /* number of variables     */

   /* 1. check range of Delta; make sure p and Delta are feasible   */
   PP = p`*p;         PQ = p`*q;
   QP = q`*p;         QQ = q`*q;
   A = -sqrt(PP/QQ);  B = -sqrt(QQ/PP);  /* matrices                */
   LBound = choose(A>B,A,B);             /* elementwise max(A or B) */
   LBound[loc(I(d))] = 1;                /* set diagonal to 1       */
   A =  sqrt(PQ/QP);  B =  sqrt(QP/PQ);
   UBound = choose(A<B,A,B);             /* min(A or B)             */
   UBound[loc(I(d))] = 1;                /* set diagonal to 1       */

   /* return 1 <==> specified  means and correlations are feasible  */
   return( all(Delta >= LBound) & all(Delta <= UBound) );
finish;
```

The implementation uses matrix arithmetic instead of DO loops to compute the bounds. The matrix PP contains products of the form $p_j p_k$, the matrix QQ contains products of the form $q_j q_k$, and so forth. The matrices A and B contain the square root of ratios of products, which are used to construct the upper and lower bounds for the elements of the correlation matrix.

9.2.2   Solving for the Intermediate Correlations

In SAS software, the PROBBNRM function computes the cumulative bivariate normal distribution. For an ordered pair (x, y), the PROBBNRM function returns the probability that an observation (X, Y) is less than or equal to (x, y), where X and Y are standard normal random variables with correlation ρ. You can use the QUANTILE function to compute the quantile of the normal distribution. Consequently, the following module evaluates the function in Step 2 of the algorithm. The parameters p_j, p_k, and δ_{jk} are passed in as global variables, where p_j is the probability of success for the jth binary variable, p_k is the probability of success for the kth binary variable, and δ_{jk} is the target correlation between the binary variables.

```
/* Objective: Find correlation, rho, that is zero of this function.
   Global variables:
   pj = prob of success for binary var Xj
   pk = prob of success for binary var Xk
   djk = target correlation between Xj and Xk    */
start MVBFunc(rho)   global(pj, pk, djk);
   Phi = probbnrm(quantile("Normal",pj), quantile("Normal",pk), rho);
   qj = 1-pj; qk = 1-pk;
   return( Phi - pj*pk - djk*sqrt(pj*qj*pk*qk) );
finish;
```

9.2.3 Simulating Multivariate Binary Variates

By using the modules that are defined in the preceding sections, you can write a function that generates multivariate binary variates with a given set of expected values and a specified correlation structure. If you are running SAS/IML 12.1 or later, then you can use the FROOT function to find the intermediate correlations. Prior to SAS/IML 12.1, you can use the Bisection module, which is included in Appendix A.

```
start RandMVBinary(N, p, Delta) global(pj, pk, djk);
   /* 1. Check parameters. Compute lower/upper bounds for all (j,k) */
   if ^CheckMVBinaryParams(LBound, UBound, p, Delta) then do;
      print "The specified correlation is invalid." LBound Delta UBound;
      STOP;
   end;

   q = 1 - p;
   d = ncol(Delta);                              /* number of variables  */

   /* 2. Construct intermediate correlation matrix by solving the
         bivariate CDF (PROBBNRM) equation for each pair of vars */
   R = I(d);
   do j = 1 to d-1;
      do k = j+1 to d;
         pj=p[j]; pk=p[k]; djk = Delta[j,k];      /* set global vars */
         *R[j,k] = bisection(LBound[j,k], UBound[j,k]);  /* pre-12.1 */
         R[j,k] = froot("MVBFunc", LBound[j,k]||UBound[j,k]);/* 12.1 */
         R[k,j] = R[j,k];
      end;
   end;

   /* 3: Generate MV normal with mean 0 and covariance R */
   X = RandNormal(N, j(1,d,0), R);
```

```
   /* 4: Obtain binary variable from normal quantile */
   do j = 1 to d;
      X[,j] = (X[,j] <= quantile("Normal", p[j]));  /* convert to 0/1 */
   end;
   return (X);
finish;
```

The RANDMVBINARY function implements the Emrich-Piedmonte algorithm. Given a vector of probabilities, `p`, and a desired correlation matrix `Delta`, the RANDMVBINARY function returns an $N \times d$ matrix that contains zeros and ones. Each column of the returned matrix is a binary variate, and the sample mean and correlation of the simulated data should be close to the specified parameters, as shown in the following example:

```
call randseed(1234);
p = {0.25 0.75 0.5};                    /* expected values of the X[j]  */
Delta = { 1    -0.1  -0.25,
          -0.1   1    0.25,
          -0.25 0.25  1   };            /* correlations between the X[j] */
X = RandMVBinary(1000, p, Delta);
/* compare sample estimates to parameters */
mean = mean(X);
corr = corr(X);
print p, mean, Delta, corr[format=best6.];
```

Figure 9.1 Parameters and Sample Estimates from Simulated Binary Data

p		
0.25	0.75	0.5

mean		
0.257	0.771	0.473

Delta		
1	-0.1	-0.25
-0.1	1	0.25
-0.25	0.25	1

corr		
1	-0.104	-0.268
-0.104	1	0.1874
-0.268	0.1874	1

The matrix **x** contains 1,000 random draws from the multivariate binary distribution with the specified expected values and correlations. Figure 9.1 shows that the mean of each column is close to the expected value of the distribution. The sample correlation between columns is close to the specified correlations between variables.

9.3   Generating Multivariate Ordinal Variates

Simulating binary correlated variables is a particular case of the more general problem of simulating multivariate correlated ordinal variables. This section describes the algorithm of Kaiser, Träger, and Leisch (2011), which builds on the work of Demirtas (2006).

In particular, this section simulates data from a multivariate distribution with $d > 2$ ordinal variables, $X_1, X_2, \ldots, X_d$. Assume that the jth variable has N_j values, $1, 2, \ldots, N_j$. Each random variable is defined by its probability mass function (PMF): $P(X_j = i) = P_{ij}$ for $i = 1, 2, \ldots, N_j$.

9.3.1   Overview of the Mean Mapping Method

The paper by Kaiser, Träger, and Leisch (2011) contains two algorithms. The algorithm presented here is called the "mean mapping method." The algorithm generalizes the Emrich-Piedmonte method that is described in Section 9.2. The algorithm is as follows:

1. (Optional) Given marginal PMFs for the ordinal variables and a target correlation matrix, Δ, that describes their pairwise correlations, check that it is feasible to solve the problem. The feasibility of the problem is not checked in this book.

2. For each pair of ordinal variables, X_i and X_j, solve a complicated equation that involves the cumulative distribution function for the standard bivariate normal distribution. The solution to the equation gives a pairwise correlation, ρ_{ij}. Let R be the matrix with $R_{ij} = \rho_{ij}$.

3. Generate multivariate normal variables $(Y_1, Y_2, \ldots, Y_k) \sim \text{MVN}(0, R)$.

4. Use quantiles of the univariate standard normal distribution to convert Y_j to X_j.

The general idea of the algorithm is shown in Figure 9.2. You generate multivariate normal data according to some correlation matrix, R. You can then use the probability of each marginal category to discretize the data. In the figure, normal bivariate points with coordinates in the lower left corner are assigned the value 1 for the first ordinal variable and the value 1 for the second variable. Points in the upper left corner are assigned 1 for the first ordinal variable and 4 for the second variable. Points near the origin are assigned to 1 or 2 for the first variable and 3 for the second, and so on.

Because of the discretization process, the correlation between the ordinal variables will usually be different from the correlation of the multivariate normal distribution from which the variables were generated. The purpose of Step 2 is to find the matrix R that gives rise to the desired correlation matrix, C.

Exercise 9.1: Let Z be an ordinal random variable with outcomes 1–4 and probability vector $\{0.2, 0.15, 0.25, 0.4\}$. Use the "Table" distribution (see Section 2.4.5) to simulate 10,000 observations from this distribution. Compute the sample mean and variance of the data.

Figure 9.2 Generate Correlated Ordinal Values from Bivariate Normal Data

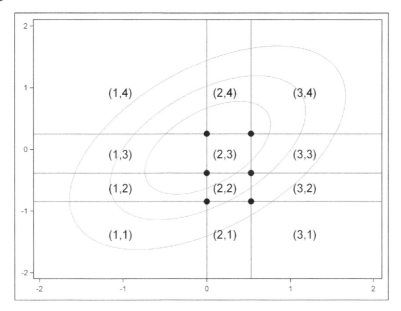

9.3.2   Simulating Multivariate Ordinal Variates

This book's Web site contains a set of SAS/IML functions for simulating multivariate ordinal data. You can download and store the functions in a SAS/IML library, and use the LOAD statement to read the modules into the active PROC IML session, as follows:

```
/* Define and store the functions for random ordinal variables */
%include "C:\<path>\RandMVOrd.sas";

proc iml;
load module=_all_;                    /* load the modules */
```

To represent a multivariate ordinal distribution, you need to specify the marginal PMFs and the desired correlations between variables. One way to represent the PMFs is to store them as columns of a matrix, P. If you have d ordinal variables and the jth variable has N_j values, then P is an $m \times d$ matrix, where $m = \max_j N_j$. You can use missing values to "pad" the matrix for variables that have a small number of values. This is the representation used by the modules in this section.

Let **P** be a SAS/IML matrix that specifies the PMFs for d ordinal variables. The following functions are used in this book:

- The ORDMEAN function, which returns the expected values for a set of ordinal random variables. The syntax is **mean = OrdMean(P)**.

- The ORDVAR function, which returns the variance for a set of ordinal random variables. The syntax is **var = OrdVar(P)**.

- The RANDMVORDINAL function, which generates N observations from a multivariate distribution of correlated ordinal variables. The syntax is **X = RandMVOrd(N, P, Corr)**, where **Corr** is the $d \times d$ correlation matrix.

For example, the following matrix contains the PMFs of three ordinal variables. The first variable has two values and the PMF {0.25, 0.75}; the last variable has four values and the PMF {0.20, 0.15, 0.25, 0.40}. The ORDMEAN and ORDVAR functions are used to display the expected values and the variances of the three random variables:

```
      /* P1    P2      P3  */
P = {0.25   0.50   0.20 ,
     0.75   0.20   0.15 ,
       .    0.30   0.25 ,
       .      .    0.40 };

/* expected values and variance for each ordinal variable */
Expected = OrdMean(P) // OrdVar(P);
varNames = "X1":"X3";
print Expected[r={"Mean" "Var"} c=varNames];
```

Figure 9.3 Expected Values and Variances for Ordinal Variables

	Expected		
	X1	X2	X3
Mean	1.75	1.8	2.85
Var	0.1875	0.76	1.3275

The main function—and the only one that you need to call explicitly to simulate ordinal data—is the RANDMVORDINAL function, which implements the mean mapping method. The following statements define a correlation matrix and generate 1,000 observations from a multivarate correlated ordinal distribution. The first few observations are shown in Figure 9.4.

```
/* test the RandMVOrd function */
Delta = {1.0   0.4   0.3,
         0.4   1.0   0.4,
         0.3   0.4   1.0 };

call randseed(54321);
X = RandMVOrdinal(1000, P, Delta);
first = X[1:5,];
print first[label="First 5 Obs: Multivariate Ordinal"];
```

Figure 9.4 A Few Simulated Observations

First 5 Obs: Multivariate Ordinal		
1	3	4
2	1	4
2	1	1
2	1	4
2	2	3

The **x** matrix is 1000×3 for this example. You can check the sample mean and sample variance, and compare them to the expected values that are shown in Figure 9.3:

```
mv = mean(X) // var(X);
corr = corr(X);
varNames = "X1":"X3";
print mv[r={"Mean" "Var"} c=varNames], corr;
```

Figure 9.5 Sample Mean and Correlation for Ordinal Data

mv			
	X1	X2	X3
Mean	1.754	1.762	2.868
Var	0.1856697	0.7501061	1.3218979

corr		
1	0.3686835	0.2960641
0.3686835	1	0.3825822
0.2960641	0.3825822	1

The sample means and variances are close to the population values. The sample correlations are close to the specified correlation matrix.

Exercise 9.2: Write the **x** matrix to a data set, and use PROC FREQ to compute the sample percentages of each category for the ordinal variables. Compare the sample percentages to the PMF for each variable.

Exercise 9.3: Use the TIME function to time how long it takes to generate a million observations for three correlated ordinal variables. The algorithm presented here is orders of magnitude faster than the algorithm in Kaiser, Träger, and Leisch (2011).

9.4 Reordering Multivariate Data: The Iman-Conover Method

The Iman-Conover method (Iman and Conover 1982) is a clever technique that combines p univariate distributions into a multivariate distribution for which the marginal distributions are *exactly* the same as the original univariate distributions. Furthermore, pairs of variables in the multivariate distribution have a rank correlation that is close to a specified target value. A rank correlation is a nonparametric measure of association that is computed by using the ranks of the data values rather than using the values themselves. The CORR procedure can compute the Spearman rank-order correlation.

The Iman-Conover method enables you to specify any data (simulated or real) for the marginal distributions. The only other input needed is the target rank correlation matrix.

Iman and Conover carefully describe each step of the construction and give an example for constructing six correlated variables. Their paper is very readable and so this book does not repeat their arguments. The main idea is as follows. Suppose M is a simulated $N \times p$ data matrix. Compute the ranks of the elements in each column of M. By definition, *any* matrix that has columns with the same ranks as M also has the same rank correlation as M. Therefore, given any $N \times p$ matrix Y, sort each column so that the column has the same ranks as the corresponding column in M. The rearranged version of Y has the specified rank correlation, but you have not changed the marginal distribution of the Y data because all you did was rearrange the columns.

The Iman-Conover article shows how to generate a matrix M that has (approximately) a given rank correlation. (This is the main contribution of the article.) There is nothing random in the Iman-Conover method. That is, given a rank correlation, the algorithm constructs the same matrix M every time. Therefore, the randomness in a simulation comes from simulating the marginal distributions of Y. The Iman-Conover algorithm simply rearranges the elements of columns of your data matrix.

The following SAS/IML program defines a function that reorders the elements in each column of a data matrix, Y, so that it has approximately the rank correlation given by the matrix C:

```
/* Use Iman-Conover method to generate MV data with known marginals
   and known rank correlation. */
proc iml;
start ImanConoverTransform(Y, C);
   X = Y;
   N = nrow(X);
   R = J(N, ncol(X));
   /* compute scores of each column */
   do i = 1 to ncol(X);
      h = quantile("Normal", rank(X[,i])/(N+1) );
      R[,i] = h;
   end;
   /* these matrices are transposes of those in Iman & Conover */
   Q = root(corr(R));
   P = root(C);
   S = solve(Q,P);                    /* same as  S = inv(Q) * P; */
   M = R*S;              /* M has rank correlation close to target C */

   /* reorder columns of X to have same ranks as M.
      In Iman-Conover (1982), the matrix is called R_B. */
   do i = 1 to ncol(M);
      rank = rank(M[,i]);
      y = X[,i];
      call sort(y);
      X[,i] = y[rank];
   end;
   return( X );
finish;
```

To test the algorithm, simulate data vectors from various distributions and pack the vectors into columns of a matrix, A, as follows:

```
/* Step 1: Specify marginal distributions */
call randseed(1);
N = 100;
A = j(N,4);    y = j(N,1);
distrib = {"Normal" "Lognormal" "Expo" "Uniform"};
do i = 1 to ncol(distrib);
   call randgen(y, distrib[i]);
   A[,i] = y;
end;
```

This example generates data from one normal distribution and three nonnormal distributions. To obtain a specified target rank correlation without changing the marginal distributions, call the Iman-Conover algorithm. Figure 9.6 shows that the rank correlation of **x** is close to the specified rank correlation matrix, C.

```
/* Step 2: specify target rank correlation */
C = { 1.00   0.75 -0.70   0,
      0.75   1.00 -0.95   0,
     -0.70  -0.95  1.00 -0.2,
      0      0     -0.2   1.0};

X = ImanConoverTransform(A, C);
RankCorr = corr(X, "Spearman");
print RankCorr[format=5.2];
```

Figure 9.6 Rank Correlation for Simulated Data with Specified Marginals

RankCorr			
1.00	0.73	-0.70	0.03
0.73	1.00	-0.95	0.03
-0.70	-0.95	1.00	-0.20
0.03	0.03	-0.20	1.00

You can visualize the simulated data to show that the marginal distributions are normally, lognormally, exponentially, and uniformly distributed, respectively. Figure 9.7 shows that the marginal distributions appear to have the specified distributions, and the pairwise correlations also appear to be as specified.

```
/* write to SAS data set */
create MVData from X[c=("x1":"x4")];  append from X;  close MVData;
quit;

proc corr data=MVData Pearson Spearman noprob plots=matrix(hist);
   var x1-x4;
run;
```

Exercise 9.4: Use PROC UNIVARIATE to fit the lognormal and exponential distributions to the x2 and x3 variables, respectively.

Figure 9.7 Univariate and Bivariate Distributions of Simulated Data

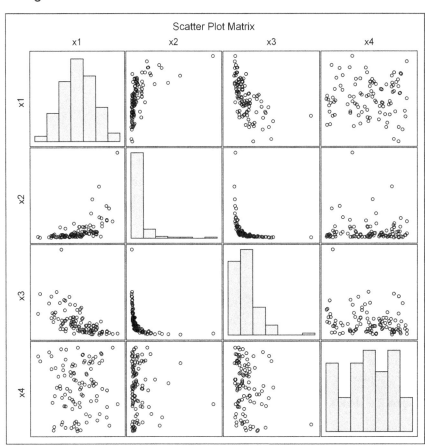

9.5   Generating Data from Copulas

Each of the previous sections describes how to combine marginal distributions (binary, ordinal, or arbitrary) to obtain a joint distribution with a specified correlation. The word "copula" means to link or join, and that is exactly what a mathematical copula does: It creates a multivariate distribution by joining univariate marginal distributions.

Copulas are mathematically sophisticated, and the copula literature is not always easy to read. However, the good news is that you can use copulas to simulate data without needing to understand all the details of their construction. This section begins with a motivating example, and then discusses some of the theory behind copulas. Section 9.5.3 describes how to use the COPULA procedure in SAS/ETS software to simulate data.

9.5.1   A Motivating Example

Suppose that you want to simulate data from a bivariate distribution that has the following properties:

- The correlation between the variables is 0.6.
- The marginal distribution of the first variable is Gamma(4) with unit scale.
- The marginal distribution of the second variable is standard exponential.

How might you accomplish this? One approach would be to try to transform a set of bivariate normal variables. It is easy to simulate multivariate normal data with a desired correlation, so perhaps you can transform the normal variates to match the desired marginal distributions. There is no reason to think that the transformed variates will have the same correlation as the normal variates, but ignore that problem for now.

Start by simulating multivariate normal data, as shown in the following SAS/IML statements:

```
proc iml;
call randseed(12345);
Sigma = {1.0  0.6,
         0.6  1.0};
Z = RandNormal(1e4, {0,0}, Sigma);
```

The matrix z contains 10,000 observations drawn from a bivariate normal distribution with correlation coefficient $\rho = 0.6$. (With 10,000 observations, the density of z closely matches the density of the MVN distribution.) To transform the columns, you can use the following basic probability fact:

Fact 1: If F is the cumulative distribution function of a continuous random variable, X, then the random variable $U = F(X)$ is distributed as $U(0, 1)$. The random variable U is called the *grade* of X.

This fact is very useful. It means that you can transform the normal variates into uniform variates by applying Φ, which is the cumulative distribution function for the univariate normal distribution. In SAS, the CDF function applies the cumulative distribution function, as follows:

```
U = cdf("Normal", Z);          /* columns of U are U(0,1) variates */
```

The columns of u are samples from a standard uniform distribution. However, they are not independent. They have correlation because they are a transformation of correlated variables.

Uniform variates are useful because they are easy to transform into *any* distribution: Just apply Fact 1 again. To make it easier to apply, rewrite Fact 1 in terms of the inverse CDF:

Fact 2: If U is a uniform random variable on $[0, 1]$ and F is the cumulative distribution function of a continuous random variable, then the random variable $X = F^{-1}(U)$ is distributed as F.

This formulation implies that you can obtain gamma variates from the first column of u by applying the inverse gamma CDF. Similarly, you can obtain exponential variates from the second column of u by applying the inverse exponential CDF. In SAS, the QUANTILE function applies the inverse CDF, as follows:

```
gamma = quantile("Gamma", U[,1], 4);     /* gamma ~ Gamma(alpha=4) */
expo = quantile("Expo", U[,2]);          /* expo ~ Exp(1)          */
X = gamma || expo;
```

At this point, you have generated correlated bivariate observations. The first column of x contains gamma variates and the second column contains exponential variates. However, it is not clear whether the variates have the desired correlation of $\rho = 0.6$. The following statements compute the correlation coefficients for the original normal variates and for the transformed variates:

```
/* if Z~MVN(0,Sigma), corr(X) is often close to Sigma,
   where X=(X1,X2,...,Xm) and X_i = F_i^{-1}(Phi(Z_i)) */
rhoZ = corr(Z)[1,2];                       /* extract corr coefficient */
rhoX = corr(X)[1,2];
print rhoZ rhoX;
```

Figure 9.8 Sample Correlation Coefficients for Normal and Transformed Variates, $\rho = 0.6$

rhoZ	rhoX
0.5983038	0.5647988

As expected, the sample correlation of the normal variates is very close to the target value of 0.6. The correlation of the transformed variates is not as close, and a two-sided 95% confidence interval for the correlation coefficient does not include 0.6 (see Exercise 9.5).

However, perhaps this approach can be modified. Suppose that you run a computer experiment to study the relationship between the correlation of the normal variates and the correlation of the transformed variates. Suppose that the experiment reveals that the sample correlation for the transformed variables is a monotone function of ρ. This implies that there is some value ρ^* such that if normal variates are drawn from a bivariate normal distribution with correlation ρ^*, then the transformed variables will have the desired target correlation of 0.6. The following statements carry out this experiment. The results are shown in Figure 9.9.

```
/* even though corr(X) ^= Sigma, you can often choose a target
   correlation, such as 0.6, and then choose Sigma so that corr(X)
   has the target correlation. */
Z0=Z; U0=U; X0=X;                          /* save original data */
Sigma = I(2);
rho = T( do(0.62, 0.68, 0.01) );
rhoTarget = j(nrow(rho), 1);
do i = 1 to nrow(rho);
   Sigma[1,2]=rho[i]; Sigma[2,1]=Sigma[1,2];
   Z = RandNormal(1e4, {0,0}, Sigma);      /* Z ~ MVN(0,Sigma) */
   U = cdf("Normal", Z);                   /* U_i ~ U(0,1)     */
   gamma = quantile("Gamma", U[,1], 4);    /* X_1 ~ Gamma(4)   */
   expo = quantile("Expo", U[,2]);         /* X_2 ~ Expo(1)    */
   X = gamma||expo;
   rhoTarget[i] = corr(X)[1,2];            /* corr(X) = ?      */
end;
print rho rhoTarget[format=6.4];
```

Figure 9.9 Sample Correlation Coefficients for Normal and Transformed Variates, $0.6 \leq \rho \leq 0.7$

rho	rhoTarget
0.62	0.5767
0.63	0.5931
0.64	0.5964
0.65	0.6128
0.66	0.6279
0.67	0.6292
0.68	0.6455

Given a target correlation $\rho_0 = 0.6$ for the transformed variables, Figure 9.9 shows that it is possible to choose a so-called *intermediate correlation*, $\rho^* \approx 0.64$, for the normal variates so that the desired correlation is achieved. The value of the intermediate correlation depends on the target value and on the specific forms of the marginal distributions.

Finding the intermediate value often requires finding the root of an equation that involves the CDF and inverse CDF. In fact, Step 2 in Section 9.2 is an example of using such an equation to generate correlated multivariate binary data.

This example generalizes to multivariate data. A copula is an abstraction of these ideas, but the example contains all of the main ideas of copulas. If you read the multivariate simulation literature, then you will see variations of this approach over and over again.

As many authors have remarked and as is shown in Section 9.4, if you use rank correlation (for example, Spearman's correlation), then the problem of finding an intermediate correlation vanishes. Because rank correlations are invariant under monotonic transformations of the data, the rank correlations of z, u, and x are equal, as shown in Figure 9.10:

```
RankCorrZ = corr(Z0, "Spearman")[1,2];
RankCorrU = corr(U0, "Spearman")[1,2];
RankCorrX = corr(X0, "Spearman")[1,2];
print RankCorrZ RankCorrU RankCorrX;
```

Figure 9.10 Rank Correlations, $\rho = 0.6$

RankCorrZ	RankCorrU	RankCorrX
0.5818607	0.5818607	0.5818607

Although the uniform variates seem to be an intermediate step, they are important because they can be used to visualize the joint dependence between the marginal distributions. In fact, it is instructive to visualize the dependencies between each of the three sets of variables: the normal variates (z), the uniform variates (u), and the transformed variates (x). The following statements write these variables to a SAS data set for further analysis:

```
Q = Z||U||X;
labels = {Z1 Z2 U1 U2 X1 X2};
create CorrData from Q[c=labels];
append from Q;
close CorrData;
```

You can use PROC CORR or PROC SGPLOT to create a scatter plot of the correlated uniform variates. The following statement creates the scatter plot, which is shown in Figure 9.11. The plot shows the copula of the joint distribution:

```
proc sgplot data=CorrData(obs=1000);
   scatter x=U1 y=U2;
run;
```

Figure 9.11 Plot of Uniform Variates Showing the Dependence Structure

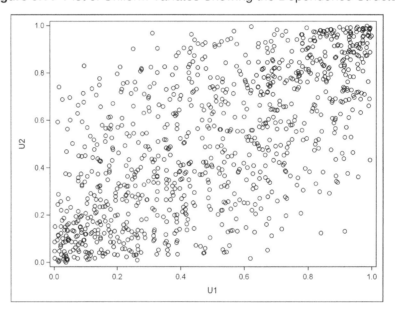

Exercise 9.5: The FISHER option in the PROC CORR statement computes a confidence interval for the correlation and can test the hypothesis that it has a specified value. Verify that the 95% two-sided confidence interval for the correlation coefficient contains 0.6 and that the null hypothesis ($\rho = 0.6$) is not rejected. Repeat the analysis for the X1 and X2 variables.

Exercise 9.6: Create a scatter plot and histograms of the X1 and X2 variables in the CorrData data set.

9.5.2   The Copula Theory

The example in the previous section is an example of a "normal copula" (Nelsen 1999). Two gentle introductions to an otherwise formidable subject are Meucci (2011) and Channouf and L'Ecuyer (2009). In certain engineering fields, the normal copula is known as the NORTA method, where NORTA means "NORmal To Anything" (Cario and Nelson 1997). This exposition follows Channouf and L'Ecuyer (2009). Although this section is written for the relatively simple case of a normal copula and continuous marginals, many of the ideas also apply to general classes of copulas and to distributions with discrete marginals.

As described in the previous section, the goal is to simulate from a joint multivariate distribution with a specified correlation and a specified set of marginal distributions. Sklar's theorem (Schmidt 2007) says that every joint distribution can be written as a function (called the *copula*) of the marginal distributions. The word "copula" means to link or join, and that is exactly what a mathematical copula does: It joins the marginal distributions in such a way as to form a joint distribution with a given correlation structure.

For completeness, here is the formal definition. Let $X = (X_1, X_2, \ldots, X_m)$ be a random vector with marginal cumulative distribution functions $F_{X_1}, F_{X_2}, \ldots, F_{X_m}$. Sklar's theorem says that the joint distribution function, F_X, can be written as a function of the marginal distributions:

$$F_X(x) = C(F_{X_1}(x_1), \ldots, F_{X_m}(x_m))$$

where $x = (x_1, \ldots, x_m)$ and $C : [0, 1]^m \to [0, 1]$ is a joint distribution function of the random variables $U_i = F_{X_i}(X_i)$.

Notice two points. First, if the X_i are independent, then the copula function is simply the product function $C(u_1, \ldots, u_m) = \Pi_i u_i$, and the corresponding density function for the copula is the constant function, 1. Second, Figure 9.11 is a graph that helps to visualize the copula function. The copula function for the example is approximated by the empirical cumulative distribution that is shown in the scatter plot.

The beauty of the copula approach is that the copula separates the problem of modeling the marginals from the problem of fitting the correlations between the variables. These two steps can be carried out independently. The copula also makes it easy to simulate data from the model: Generate correlated uniform variates according to the copula (as in Figure 9.11), and then generate the random variates for each coordinate by applying the appropriate inverse distribution function: $X_i = F_{X_i}^{-1}(U_i)$. The generation of the random uniform variates depends on the choice of the copula, but for a normal copula you can use the technique in Section 9.5.1.

9.5.3 Fitting and Simulating Data from a Copula Model

In SAS software, you can use the MODEL procedure or the COPULA procedure in SAS/ETS software to fit a copula to data and to simulate from the copula (Erdman and Sinko 2008; Chvosta, Erdman, and Little 2011). This section uses the COPULA procedure to fit a normal copula model and to simulate data with similar characteristics. For more complicated examples, see the cited papers and the documentation for the COPULA procedure in the *SAS/ETS User's Guide*.

Rank (Spearman) correlations, rather than Pearson correlations, are used by PROC COPULA and by many who use copulas. In copula theory, there are three primary kinds of dependence structures (Schmidt 2007): Pearson correlation, rank correlation, and tail dependence. The classical Pearson correlation is a suitable measure of dependence only for the so-called "elliptical distributions," which includes normal copulas, Student t copulas, and their mixtures. Outside of this class of distributions, Pearson correlations are unwieldy to work with because of the following shortcomings (Schmidt 2007, p. 14):

- Pearson correlations are not invariant under general monotonic transformation.

- As was shown for multivariate binary variables in Section 9.2, for specified marginal distributions there might be Pearson correlations that are unattainable.

In contrast, rank correlations do not suffer from these shortcomings.

The copula process proceeds in four steps:

1. Model the marginal distributions.

2. Choose a copula from among those supported by PROC COPULA. Fit the copula to the data to estimate the copula parameters. For this example, a normal copula is used, so the parameters are the six pairwise correlation coefficients that make up the upper-triangular correlation matrix of the data.

3. Simulate from the copula. For the normal copula, this consists of generating multivariate normal data with the given rank correlations. These simulated data are transformed to uniformity by applying the normal CDF to each component.

4. Transform the uniform marginals into the marginal distributions by applying the inverse CDF for each component.

For this example, use the MVData data set that is described in Section 9.4. Assume that Step 1 is complete and that the following models describe each marginal distribution:

- The X1 variable is modeled by a standard normal distribution.

- The X2 variable is modeled by a standard lognormal distribution.

- The X3 variable is modeled by a standard exponential distribution.

- The X4 variable is modeled by a uniform distribution on $[0, 1]$.

Steps 2 and 3 are accomplished by using the COPULA procedure. Step 2 is accomplished with the FIT statement and Step 3 is accomplished with the SIMULATE statement, as shown in the following call to PROC COPULA:

```
/* Step 2: fit normal copula
   Step 3: simulate data, transformed to uniformity */
proc copula data=MVData;
   var x1-x4;
   fit normal;
   simulate / seed=1234  ndraws=100
              marginals=empirical  outuniform=UnifData;
run;
```

The UnifData data set contains the transformed simulated data. You can apply the final transformation by using the QUANTILE function for each component to recover the modeled form of the marginals:

```
/* Step 4: use inverse CDF to invert uniform marginals */
data Sim;
set UnifData;
normal = quantile("Normal", x1);
lognormal = quantile("LogNormal", x2);
expo = quantile("Exponential", x3);
uniform = x4;
run;
```

You can use PROC CORR to compute the rank correlations of the original and simulated data. You can also create a matrix of scatter plots that display the pairwise dependencies between the original and the simulated data.

```
/* Compare original distribution of data to simulated data */
proc corr data=MVData Spearman noprob plots=matrix(hist);
   title "Original Data";
   var x1-x4;
run;

proc corr data=Sim Spearman noprob plots=matrix(hist);
   title "Simulated Data";
   var normal lognormal expo uniform;
run;
```

Figure 9.12 Original Data

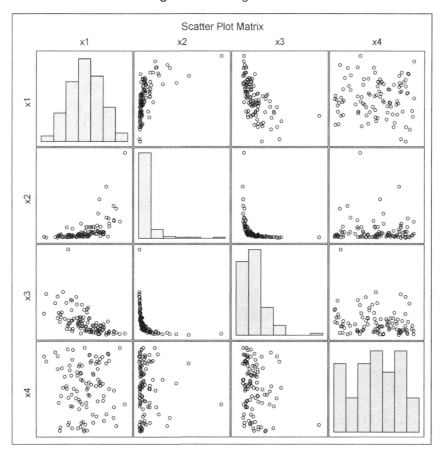

Figure 9.13 Data Simulated from a Copula

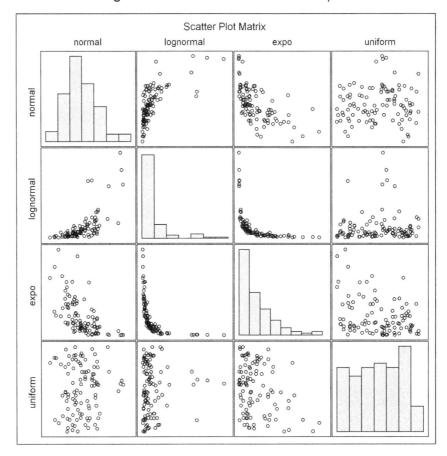

The Spearman rank correlation matrices are not shown, but the Spearman correlations of the variables in the Sim data set are close to those in the MVData data set. The scatter plot matrices are shown in Figure 9.12 and Figure 9.13. They look very similar. Unfortunately, PROC COPULA does not provide any goodness-of-fit statistics or an indication of whether the copula model is a good model for the data.

You can simulate many samples from the COPULA function by using a single call. For example, to simulate 100 samples of size 20, use the %SYSEVALF macro to generate the product of these numbers:

```
%let N = 20;
%let NumSamples=100;
proc copula data=MVData;
   ...
   simulate / ndraws=%sysevalf(&N*&NumSamples) outuniform=UnifData;
   ...
```

You can then use a separate DATA step to assign values for the usual SampleID variable, as follows:

```
data UnifData;
set UnifData;
SampleID = 1 + floor((_N_-1) / &N);
run;
```

Exercise 9.7: Use the FISHER option in the PROC CORR statement to compute 95% confidence intervals for the Spearman correlation of the variables in the Sim data set. Show that these confidence intervals contain the parameter values in Section 9.4 that generated the data.

9.6   References

Cario, M. C. and Nelson, B. L. (1997), *Modeling and Generating Random Vectors with Arbitrary Marginal Distributions and Correlation Matrix*, Technical report, Northwestern University.

Channouf, N. and L'Ecuyer, P. (2009), "Fitting a Normal Copula for a Multivariate Distribution with Both Discrete and Continuous Marginals," in M. D. Rossetti, R. R. Hill, B. Johansson, A. Dunkin, and R. G. Ingalls, eds., *Proceedings of the 2009 Winter Simulation Conference*, 352–358, Piscataway, NJ: Institute of Electrical and Electronics Engineers.

Chvosta, J., Erdman, D., and Little, M. (2011), "Modeling Financial Risk Factor Correlation with the COPULA Procedure," in *Proceedings of the SAS Global Forum 2011 Conference*, Cary, NC: SAS Institute Inc.
URL http://support.sas.com/resources/papers/proceedings11/340-2011.pdf

Demirtas, H. (2006), "A Method for Multivariate Ordinal Data Generation Given Marginal Distributions and Correlations," *Journal of Statistical Computation and Simulation*, 76, 1017–1025.

Emrich, L. J. and Piedmonte, M. R. (1991), "A Method for Generating High-Dimensional Multivariate Binary Variables," *American Statistician*, 45, 302–304.

Erdman, D. and Sinko, A. (2008), "Using Copulas to Model Dependency Structures in Econometrics," in *Proceedings of the SAS Global Forum 2008 Conference*, Cary, NC: SAS Institute Inc.
URL http://support.sas.com/resources/papers/sgf2008/copulas.pdf

Iman, R. L. and Conover, W. J. (1982), "A Distribution-Free Approach to Inducing Rank Correlation among Input Variables," *Communications in Statistics—Simulation and Computation*, 11, 311–334.

Kaiser, S., Träger, D., and Leisch, F. (2011), *Generating Correlated Ordinal Random Values*, Technical report, University of Munich, Department of Statistics.
URL http://epub.ub.uni-muenchen.de/12157/

Meucci, A. (2011), "A Short, Comprehensive, Practical Guide to Copulas," Social Science Research Network Working Paper Series.
URL http://ssrn.com/abstract=1847864

Nelsen, R. B. (1999), *An Introduction to Copulas*, New York: Springer-Verlag.

Oman, S. D. (2009), "Easily Simulated Multivariate Binary Distributions with Given Positive and Negative Correlations," *Computational Statistics and Data Analysis*, 53, 999–1005.

Schabenberger, O. and Gotway, C. A. (2005), *Statistical Methods for Spatial Data Analysis*, Boca Raton, FL: Chapman & Hall/CRC.

Schmidt, T. (2007), "Coping with Copulas," in J. Rank, ed., *Copulas: From Theory to Application in Finance*, 3–34, London: Risk Books.

Chapter 10

Building Correlation and Covariance Matrices

Contents

10.1 Overview of Building Correlation and Covariance Matrices

Most of this book is concerned with simulating data. This chapter is different. It describes how to construct correlation and covariance matrices that you can use in the algorithms that are included in other chapters. This chapter addresses three main issues:

- How to construct covariance matrices that have a particular structure.

- How to simulate random matrices that have a particular set of eigenvalues.

- How to find the correlation matrix that is closest to an arbitrary matrix.

Covariance matrices with known structure are important in many applications and are useful for simulating correlated data. Chapter 8, "Simulating Data from Basic Multivariate Distributions," includes many examples of simulating correlated data. For each example, you must specify a covariance matrix. Furthermore, correlated data are important in mixed models, time-series models, and spatial models.

Simulating random matrices can be useful for testing algorithms. The second part of this chapter introduces the Wishart distribution, which is the distribution of the sample covariance matrix of the multivariate normal distribution. Section 10.6 shows how to simulate correlation matrices that have specified eigenvalues.

The third part of this chapter discusses what to do when your estimate for a correlation matrix is poor. It might happen that an estimate does not satisfy the mathematical properties that are required for a correlation matrix. Section 10.8 describes one way to adjust your estimate.

All of the correlation matrices in this chapter are Pearson correlations. You can use the Iman-Conover algorithm in Section 9.4 to generate data with a given Spearman rank correlation.

This chapter requires more linear algebra than the rest of the book. The topics are not used heavily in subsequent chapters, so you can skip this material if matrices are not your passion. However, readers who are interested in simulating mixed models should read the chapter through Section 10.4.4.

10.2   Converting between Correlation and Covariance Matrices

Both covariance matrices and correlation matrices are used frequently in multivariate statistics. If S is a covariance matrix, then the corresponding correlation matrix is $R = D^{-1} S D^{-1}$, where D is the diagonal matrix that contains the square root of the diagonal element of S. The diagonal elements of S are the variances of the variables, so D contains the standard deviations of the variables. Although you can use this formula to convert a covariance matrix to a correlation matrix, the following SAS/IML function is an equivalent formulation that is computationally more efficient:

```
proc iml;
/* convert a covariance matrix, S, to a correlation matrix */
start Cov2Corr(S);
   D = sqrt(vecdiag(S));
   return( S / D` / D );          /* divide columns, then divide rows */
finish;
```

Rather than perform matrix inversion and matrix multiplication, the function divides each column of S by the corresponding element of D, and then divides each row in the same way.

In a similar way, you can convert a correlation matrix into a covariance matrix provided that you know the standard deviations of each variable: $S = DRD$. The following function is computationally equivalent to the matrix multiplication, but uses scalar multiplication of columns and rows, which is more efficient:

```
/* R = correlation matrix
   sd = (vector of) standard deviations for each variable
   Return covariance matrix with sd##2 on the diagonal */
start Corr2Cov(R, sd);
   std = colvec(sd);                /* convert to a column vector */
   return( std` # R # std );
finish;
```

The following statements test these functions. First, define a covariance matrix and convert it to a correlation matrix. Then use the standard deviation of each variable to convert it back to a covariance matrix. Figure 10.1 shows that the process recovers the original covariance matrix.

```
S = {1.0  1.0  8.1,              /* covariance matrix            */
     1.0 16.0 18.0,
     8.1 18.0 81.0 };
Corr = Cov2Corr(S);              /* convert to correlation matrix */

sd = sqrt(vecdiag(S));           /* sd = {1 4 9}                 */
Cov = Corr2Cov(Corr, sd);        /* convert to covariance matrix */
print Corr, Cov;
```

Figure 10.1 Correlation and Covariance Matrices

Corr		
1	0.25	0.9
0.25	1	0.5
0.9	0.5	1

Cov		
1	1	8.1
1	16	18
8.1	18	81

In this chapter, some algorithms use a covariance matrix whereas others use a correlation matrix. If necessary, convert the matrix that you have into the form that is required by the algorithm.

10.3   Testing Whether a Matrix Is a Covariance Matrix

When simulating multivariate data, you typically need to provide a covariance (or correlation) matrix to an algorithm that generates the random samples. This is sometimes difficult because not every matrix is a valid covariance matrix. The matrix must be symmetric and *positive definite* (PD). A positive definite matrix is a matrix that has all positive eigenvalues. A *positive semidefinite* (PSD) matrix is a matrix that has nonnegative eigenvalues.

Being PSD is equivalent to being a covariance matrix. Every covariance matrix is PSD, and every symmetric PSD matrix is a covariance matrix for some distribution. For brevity, this chapter typically refers to positive definite matrices, although most results and algorithms are also valid for positive semidefinite matrices.

Suppose that you want to test whether a given matrix is a valid covariance matrix. You have to check two properties: that the matrix is symmetric, and that the matrix is positive semidefinite.

10.3.1   Checking Whether a Matrix Is Symmetric

The first step is to check whether the matrix is symmetric. For a small matrix, you can verify symmetry "by eye." However, for a 100×100 matrix, it is better to have the computer check symmetry for you.

Mathematically, a matrix A is symmetric if $B = A$, where $B = (A + A')/2$. However, for finite-precision computations it only makes sense to ask whether $\max_{ij} |B_{ij} - A_{ij}|$ is small. The following SAS/IML program checks that the matrix is symmetric to within the scale of the data:

```
proc iml;
A = { 2 -1  0,
     -1  2 -1,
      0 -1  2 };

/* finite-precision test of whether a matrix is symmetric */
start SymCheck(A);
   B = (A + A`)/2;
   scale = max(abs(A));
   delta = scale * constant("SQRTMACEPS");
   return( all( abs(B-A)< delta ) );
finish;

/* test a matrix for symmetry */
IsSym = SymCheck(A);
print IsSym;
```

Figure 10.2 Check for Symmetry

IsSym
1

Figure 10.2 shows that the SYMCHECK function returns the value 1, which means that the matrix A is symmetric to within numerical precision. The SYMCHECK function uses the CONSTANT function in Base SAS to compute the square root of *machine precision* (also known as *machine epsilon*), which is a quantity that is used in numerical analysis to understand the relative error of finite-precision computations.

10.3.2   Checking Whether a Matrix Is Positive Semidefinite

The second step is to check whether a matrix is positive semidefinite. You can use the following SAS/IML functions to test whether a symmetric matrix, A, is PSD:

1. Use the ROOT function to compute a Cholesky decomposition. If the decomposition succeeds, the matrix is PSD.

2. Use the EIGVAL function to compute the eigenvalues of *A*. If no eigenvalue is negative, *A* is PSD. If all eigenvalues are positive, *A* is positive definite.

The ROOT function is faster than the EIGVAL function, so use the ROOT function when you just need a yes-or-no answer, as follows:

```
G = root(A);
```

If the matrix is not PSD, the ROOT function will stop with an error: `ERROR: Matrix should be positive definite`. If the ROOT function does not stop with an error, then the matrix is a valid covariance matrix.

In SAS/IML 12.1 and later, the ROOT function supports an optional second parameter that you can use to suppress the error message. With this syntax, the ROOT function never stops with an ERROR. If *A* is PSD, then the return value is the Cholesky root of *A*. Otherwise, the return value is a matrix of missing values. An example follows:

```
G = root(A, "NoError");                    /* SAS/IML 12.1 and later */
if G=. then print "The matrix is not positive semidefinite";
```

Although the EIGVAL function is not as fast as the ROOT function, the EIGVAL function has the advantage that you can inspect the eigenvalues. In particular, the magnitude of the smallest eigenvalue gives information about how close the matrix is to being PSD. A small negative eigenvalue means that the matrix is close to being semidefinite.

The following statements compute and print the eigenvalues for the example matrix in this section. If any eigenvalue is negative, the matrix is not PSD.

```
eigval = eigval(A);
print eigval;
if any(eigval<0) then print "The matrix is not positive semidefinite";
```

Figure 10.3 Eigenvalues of a Positive Definite Matrix

eigval
3.4142136
2
0.5857864

10.4   Techniques to Build a Covariance Matrix

Positive definite matrices are special. You cannot write down an arbitrary symmetric matrix and expect it to be PD. This section describes the following techniques for generating a PD covariance or correlation matrix:

- Estimate a covariance matrix from real or simulated data.

- Generate a symmetric matrix that is *diagonally dominant*.

- Generate a matrix that has a special structure that is known to be PD.

10.4.1   Estimate a Covariance Matrix from Data

If you are trying to simulate data that are similar to a set of real data, it makes sense to use the sample covariance or correlation matrix in the simulation. For example, if you have data that are approximately multivariate normal (MVN), then you can use the estimated covariance matrix in place of the (unknown) population covariance to simulate data.

Two kinds of correlation matrices that arise in practice are the correlation between variables and the correlation between observations.

You can use the CORR procedure to estimate an unstructured empirical correlation of *variables*. (More sophisticated modelers might use structural equation modeling, which is implemented in the CALIS procedure in SAS/STAT software.) The CORR procedure in Base SAS and the COV function in SAS/IML software can be used to compute covariance matrices for continuous variables. For example, the following statements show two equivalent ways to compute the covariance between numerical variables in the Sashelp.Class data set:

```
/* Method 1: Base SAS approach */
proc corr data=Sashelp.Class COV NOMISS outp=Pearson;
   var Age Height Weight;
   ods select Cov;
run;

/* Method 2: equivalent SAS/IML computation */
proc iml;
use Sashelp.Class;
read all var {"Age" "Height" "Weight"} into X;
close Sashelp.Class;

Cov = cov(X);
```

The computations in PROC CORR and PROC IML are equivalent. A covariance matrix that is computed from a data matrix, X, is symmetric positive semidefinite if the following conditions are true:

- No column of X is a linear combination of other columns.

- The number of rows of X exceeds the number of columns.

- If an observation contains a missing value in any variable, then the observation is not used to form the covariance matrix. The NOMISS option in the PROC CORR statement ensures that this condition is satisfied.

In the preceding program, the COV option in the PROC CORR statement is used to generate the covariance matrix. Furthermore, the OUTP= option writes the covariance matrix (and the correlation matrix) to a SAS data set.

In longitudinal studies, *observations* for the same subject are often correlated. These correlation matrices can be quite large, although they are often assumed to have block form. SAS software supports estimating covariance matrices that have particular structures that arise in repeated measures analyses. The MIXED (or GLIMMIX) procedure can be used for this kind of estimation. See the examples in Section 12.3.

10.4.2   Generating a Symmetric Matrix

Covariance and correlation matrices are symmetric. The first step in generating a random covariance matrix is to know how to generate a symmetric matrix. One way is to generate an arbitrary matrix A and define $M = c(A^T + A)$ for any value of c. However, this method usually changes the distributional properties of the elements. A more efficient approach is to generate the lower triangular portion of a matrix, and then use the SQRVECH function in SAS/IML software to create a full symmetric matrix. (SAS/IML also supports the VECH function, which extracts the lower triangular portion of a square matrix.)

The following SAS/IML program creates a random symmetric matrix where each element is uniformly distributed:

```
proc iml;
N = 4;                              /* want 4x4 symmetric matrix     */
call randseed(1);
v = j(N*(N+1)/2, 1);                /* allocate lower triangular     */
call randgen(v, "Uniform");         /* fill with random             */
x = sqrvech(v);                     /* create symmetric matrix from v */
print x[format=5.3];
```

Figure 10.4 A Random Symmetric Matrix

x			
0.884	0.974	0.508	0.887
0.974	0.689	0.943	0.929
0.508	0.943	0.488	0.780
0.887	0.929	0.780	0.877

The elements of the matrix **x** are distributed uniformly; change the argument to the RANDGEN function to simulate elements from other distributions.

10.4.3   Generating a Diagonally Dominant Covariance Matrix

In a simulation study, you might need to generate a random unstructured covariance matrix. For large matrices, if you generate a matrix with unit diagonal for which each off-diagonal element is uniformly and independently generated in $[-1, 1]$, chances are that the matrix is not positive definite. Although every 2×2 matrix of this form is a correlation matrix, for 3×3 matrices the probability of being PD is 61.7% and for 4×4 matrices the probability of being PD drops to 18.3% (Rousseeuw and Molenberghs 1994).

There is a useful fact for creating a PD matrix: A symmetric matrix with positive diagonal entries is PD if it is *row diagonally dominant*. A matrix A is row diagonally dominant if $|A_{ii}| > \Sigma_{j \neq i} |A_{ij}|$ for each row i. In other words, the magnitude of the diagonal element is greater than the sum of the magnitudes of the off-diagonal elements in the same row.

This means that you can generate an PD matrix by doing the following:

1. Create a random symmetric matrix as shown in Section 10.4.2.

2. For each row, increase the diagonal element until it is larger than the sum of the magnitudes of the other entries.

You can increase the diagonal element of each row independently, but a popular alternative is to find a positive value, λ, such that $B = A + \lambda \text{diag}(A)$ is positive semidefinite. If you define $s_i = \Sigma_{j \neq i} |A_{ij}|$ to be the sum of the off-diagonal elements, then B is positive semidefinite provided that $\lambda \geq \max(s_i/d_i) - 1$, where d_i is the ith diagonal element. The following SAS/IML module implements this method. The example uses the symmetric 4×4 matrix **x** in Figure 10.4.

```
/* Add a multiple of diag(A) so that A is diagonally dominant. */
start Ridge(A, scale);          /* Input scale >= 1              */
   d = vecdiag(A);
   s = abs(A)[,+] - d;          /* sum(abs of off-diagonal elements) */
   lambda = scale * (max(s/d) - 1);
   return( A + lambda*diag(d) );
finish;

/* assume x is symmetric matrix */
H = Ridge(x, 1.01);             /* scale > 1 ==> H is pos definite  */
print H;
```

Figure 10.5 A Diagonally Dominant Symmetric Matrix

H			
4.0746613	0.9738211	0.5075826	0.8869405
0.9738211	3.1779967	0.9432546	0.9287864
0.5075826	0.9432546	2.2481529	0.7798851
0.8869405	0.9287864	0.7798851	4.0436477

If the **scale** parameter equals unity, then H is PD. If the parameter is greater than 1, H is positive definite. The matrix H, which is shown in Figure 10.5, is symmetric positive definite, and therefore is a valid covariance matrix. You can confirm that a matrix is PD by using the EIGVAL function to show that the eigenvalues are positive, or by using the ROOT function to compute a Cholesky decomposition.

Exercise 10.1: Write a SAS/IML function that uses the technique in this section to generate N random covariance matrices, where N is a parameter to the function.

Exercise 10.2: Another approach is to define $B = A + \lambda(I)$. Write a SAS/IML module that takes a matrix A and computes the value of λ so that $A + \lambda(I)$ is diagonally dominant.

10.4.4   Generating a Covariance Matrix with a Known Structure

There are several well-known covariance structures that are used to model the random effects in mixed models. If you simulate data that are appropriate for a mixed model analysis, it is important to be able to generate covariance matrices with an appropriate structure. Often these structures are used to generate correlated errors from a multivariate normal distribution. See Section 12.3.

This section presents four common covariance structures that are used in mixed modeling and that are supported by the MIXED procedure in SAS: a diagonal structure, compound symmetry, Toeplitz, and first-order autoregressive (AR(1)). Examples of the covariance structures are shown in the following table:

Table 10.1   Covariance Structure Examples

Structure	Example
Variance components	$\begin{bmatrix} \sigma_1^2 & 0 & 0 & 0 \\ 0 & \sigma_2^2 & 0 & 0 \\ 0 & 0 & \sigma_3^2 & 0 \\ 0 & 0 & 0 & \sigma_4^2 \end{bmatrix}$
Compound symmetry	$\begin{bmatrix} \sigma^2 + \sigma_1 & \sigma_1 & \sigma_1 & \sigma_1 \\ \sigma_1 & \sigma^2 + \sigma_1 & \sigma_1 & \sigma_1 \\ \sigma_1 & \sigma_1 & \sigma^2 + \sigma_1 & \sigma_1 \\ \sigma_1 & \sigma_1 & \sigma_1 & \sigma^2 + \sigma_1 \end{bmatrix}$
Toeplitz	$\begin{bmatrix} \sigma^2 & \sigma_1 & \sigma_2 & \sigma_3 \\ \sigma_1 & \sigma^2 & \sigma_1 & \sigma_2 \\ \sigma_2 & \sigma_1 & \sigma^2 & \sigma_1 \\ \sigma_3 & \sigma_2 & \sigma_1 & \sigma^2 \end{bmatrix}$
First-order autoregressive	$\sigma^2 \begin{bmatrix} 1 & \rho & \rho^2 & \rho^3 \\ \rho & 1 & \rho & \rho^2 \\ \rho^2 & \rho & 1 & \rho \\ \rho^3 & \rho^2 & \rho & 1 \end{bmatrix}$

10.4.4.1   A Covariance Matrix with a Diagonal Structure

The diagonal covariance matrix is known as the *variance components* model. It is easy to create such a covariance structure in SAS/IML software, as demonstrated by the following module definition. The matrix shown in Figure 10.6 is a covariance matrix that contains specified variances along the diagonal. As long as the diagonal elements are positive, the resulting matrix is a covariance matrix.

```
proc iml;
/* variance components: diag({var1, var2,..,varN}), var_i>0 */
start VarComp(v);
   return( diag(v) );
finish;

vc = VarComp({16,9,4,1});
print vc;
```

Figure 10.6 A Covariance Matrix with Diagonal Structure

vc			
16	0	0	0
0	9	0	0
0	0	4	0
0	0	0	1

10.4.4.2 A Covariance Matrix with Compound Symmetry

The *compound symmetry* model is the sum of a constant matrix and a diagonal matrix. The compound symmetry structure arises naturally with nested random effects. The following module generates a matrix with compound symmetry. The matrix shown in Figure 10.7 is a compound symmetric matrix. To guarantee that a compound symmetric matrix is PD, choose $v > 0$ and $v_1 > -v/N$.

```
/* compound symmetry, v>0:
   {v+v1   v1    v1    v1,
      v1  v+v1    v1    v1,
      v1    v1  v+v1    v1,
      v1    v1    v1  v+v1  };
*/
start CompSym(N, v, v1);
   return( j(N,N,v1) + diag( j(N,1,v) ) );
finish;

cs = CompSym(4, 4, 1);
print cs;
```

Figure 10.7 A Covariance Matrix with Compound Symmetry Structure

cs			
5	1	1	1
1	5	1	1
1	1	5	1
1	1	1	5

The compound symmetry structure is of the form $v\mathbf{I} + v_1\mathbf{J}$, where $\mathbf{I}$ is the identity matrix and $\mathbf{J}$ is the matrix of all ones. A special case of the compound symmetry structure is the *uniform correlation* structure, which is a matrix of the form $(1 - v_1)\mathbf{I} + v_1\mathbf{J}$. The uniform correlation structure (sometimes called a constant correlation structure) is PD when $v_1 > -1/(N - 1)$.

Exercise 10.3: Write a SAS/IML function that constructs a matrix with uniform correlation structure.

10.4.4.3   A Covariance Matrix with Toeplitz Structure

A Toeplitz matrix has a banded structure. The diagonals that are parallel to the main diagonal are constant. The SAS/IML language has a built-in TOEPLITZ function that returns a Toeplitz matrix, as shown in the following example. The matrix shown in Figure 10.8 has a Toeplitz structure. Other than diagonal dominance, there is no simple rule that tells you how to choose the Toeplitz parameters in order to ensure a positive definite matrix.

```
/* Toeplitz:
   {s##2 s1   s2    s3,
    s1    s##2 s1    s2,
    s2    s1    s##2 s1,
    s3    s2    s1    s##2 };
   Let u = {s1 s2 s3};
*/
toep = toeplitz( {4 1 2 3} );
print toep;
```

Figure 10.8 A Covariance Matrix with Toeplitz Structure

toep			
4	1	2	3
1	4	1	2
2	1	4	1
3	2	1	4

10.4.4.4   A Covariance Matrix with First-Order Autoregressive Structure

A first-order autoregressive (AR(1)) structure is a Toeplitz matrix with additional structure. Whereas an $n \times n$ Toeplitz matrix has n parameters, an AR(1) structure has two parameters. The values along each diagonal are related to each other by a multiplicative factor. The following module generates a matrix with AR(1) structure by calling the module in the previous section. The matrix shown in Figure 10.9 has an AR(1) structure. An AR(1) matrix is PD when $|\rho| < 1$.

```
/* AR1 is special case of Toeplitz */
/* autoregressive(1):
   s##2 * {1        rho    rho##2 rho##3,
           rho    1        rho    rho##2,
           rho##2 rho    1        rho    ,
           rho##3 rho##2 rho    1        };
   Let u = {rho rho##2 rho##3}
*/
start AR1(N, s, rho);
   u = cuprod(j(1,N-1,rho));                  /* cumulative product */
   return( s##2 # toeplitz(1 || u) );
finish;

ar1 = AR1(4, 1, 0.25);
print ar1;
```

Figure 10.9 A Covariance Matrix with AR(1) Structure

ar1			
1	0.25	0.0625	0.015625
0.25	1	0.25	0.0625
0.0625	0.25	1	0.25
0.015625	0.0625	0.25	1

10.5   Generating Matrices from the Wishart Distribution

If you draw N observations from a MVN$(0,\Sigma)$ distribution and compute the sample covariance matrix, S, then the sample covariance is not exactly equal to the population covariance, Σ. Like all statistics, S has a distribution. The sampling distribution depends on Σ and on the sample size, N.

The Wishart distribution is the sampling distribution of $A = (N-1)S$. Equivalently, you can say that A is distributed as a Wishart distribution with $N-1$ degrees of freedom, which is written $W(\Sigma, N-1)$. Notice that $S = (X - \bar{x})^T (X - \bar{x})/(N-1)$, where X is a sample from MVN$(0,\Sigma)$ and where $\bar{x}$ is the sample mean. Therefore, A is the *scatter matrix* $(X - \bar{x})^T (X - \bar{x})$. You can simulate draws from the Wishart distribution by calling the RANDWISHART function in SAS/IML software, as shown in the following program. Two covariance matrices are shown in Figure 10.10.

```
proc iml;
call randseed(12345);
NumSamples = 1000;                    /* number of Wishart draws      */
N = 50;                               /* MVN sample size              */
Sigma = {9 1,
         1 1};
/* Simulate matrices. Each row is scatter matrix */
A = RandWishart(NumSamples, N-1, Sigma);
B = A / (N-1);                        /* each row is covariance matrix   */

S1 = shape(B[1,], 2, 2);             /* first row, reshape into 2 x 2   */
S2 = shape(B[2,], 2, 2);             /* second row, reshape into 2 x 2 */
print S1 S2;                          /* two 2 x 2 covariance matrices   */

SampleMean = shape(B[:,], 2, 2);   /* mean covariance matrix         */
print SampleMean;
```

Figure 10.10 Sample Covariance Matrices and Mean Covariance

S1		S2	
11.253995	1.3698336	8.7377434	1.3987418
1.3698336	1.1181818	1.3987418	1.1040537

Figure 10.10 *continued*

SampleMean	
8.9982748	1.01121
1.01121	0.9975039

Each row of **A** is the scatter matrix for a sample of 50 observations drawn from a MVN(0, Σ) distribution. Each row of **B** is the associated covariance matrix. The SHAPE function is used to convert a row of **B** into a 2×2 matrix. Figure 10.10 shows that the sample variances and covariance are close to the population values, but there is considerable variance in those values due to the small sample, $N = 50$. If you compute the average of the scatter matrices that are contained in **B**, you obtain the Monte Carlo estimate of Σ.

Exercise 10.4: Modify the program in this section to write the columns of **B** to a SAS data set. Use PROC UNIVARIATE to draw histograms of B_{11}, B_{12}, and B_{22}, where B_{ij} is the (i, j)th element of B. Confirm that the distribution of the estimates are centered around the population parameters. Compare the standard deviations of the diagonal elements of B (the variances) and the off-diagonal element (the covariance).

10.6 Generating Random Correlation Matrices

The "structured" covariance matrices in Section 10.4.4 are useful for generating correlated observations that arise in repeated measures analysis and mixed models. Another useful matrix structure involves eigenvalues. The set of eigenvalues for a matrix is called its *spectrum*. This section describes how to generate a random correlation matrix with a given spectrum.

The ability to generate a correlation matrix with a specific spectrum is useful in many contexts. For example, in principal component analysis, the k largest eigenvalues of the correlation matrix determine the proportion of the (standardized) variance that is explained by the first k principal components. By specifying the eigenvalues, you can simulate data for which the first k principal components explain a certain percentage of the variance. Another example arises in regression analysis where the ratio of the largest eigenvalue to the other eigenvalues is important in collinearity diagnostics (Belsley, Kuh, and Welsch 1980).

The eigenvalues of an $n \times n$ correlation matrix are real, nonnegative, and sum to n because the trace of a matrix is the sum of its eigenvalues. Given those constraints, Davies and Higham (2000) showed how to create a random correlation matrix with a specified spectrum. The algorithm consists of the following steps:

1. Generate a random matrix with the specified eigenvalues. This step requires generating a random orthogonal matrix (Stewart 1980).

2. Apply Givens rotations to convert the random matrix to a correlation matrix without changing the eigenvalues (Bendel and Mickey 1978).

This book's Web site contains a set of SAS/IML functions for generating random correlation matrices. Some of the SAS/IML programs are based on MATLAB functions written by Higham (1991) or GAUSS functions written by Rapuch and Roncalli (2001). You can download and store the functions in a SAS/IML library and use the LOAD statement to read the modules into the active PROC IML session, as follows:

```
/* Define and store the functions for random correlation matrices */
%include "C:\<path>\RandCorr.sas";

proc iml;
load module=_all_;                      /* load the modules */
```

The main function is the RANDCORR function, which returns a random correlation matrix with a given set of eigenvalues. The syntax is `R = RandCorr(lambda)`, where `lambda` is a vector with d elements that specifies the eigenvalues. If the elements of `lambda` do not sum to d, the vector is scaled appropriately.

The following statements call the RANDCORR function to create a random correlation matrix with a given set of eigenvalues. The EIGVAL function is called to verify the eigenvalues are correct:

```
/* test it: generate 4 x 4 matrix with given spectrum */
call randseed(4321);
lambda = {2 1 0.75 0.25};               /* notice that sum(lambda) = 4  */
R = RandCorr(lambda);                   /* R has lambda for eigenvalues */
eigvalR = eigval(R);                    /* verify eigenvalues           */
print R, eigvalR;
```

Figure 10.11 Random Correlation Matrix with Specified Eigenvalues

R			
1	0.2685597	0.0882014	-0.460432
0.2685597	1	0.0258407	-0.717097
0.0882014	0.0258407	1	-0.075523
-0.460432	-0.717097	-0.075523	1

eigvalR
2
1
0.75
0.25

There are many ways to use the random correlation matrix in simulation studies. For example, you can use it to create a TYPE=CORR data set to use as input to a SAS procedure. Or you can use the matrix as an argument to the RANDNORMAL function, as in Section 8.3.1. Or you can use R directly to generate multivariate correlated data, as shown in Section 8.8.

Uncorrelated data have a correlation matrix for which all eigenvalues are close to unity. Highly correlated data have one large eigenvalue and many smaller eigenvalues. There are many possibilities in between.

Exercise 10.5: Simulate 1,000 random 3×3 correlation matrices for the vector $\lambda = (1.5, 1, 0.5)$. Draw histograms for the distribution of the off-diagonal correlations.

10.7 When Is a Correlation Matrix Not a Correlation Matrix?

A problem faced by practicing statisticians, especially in economics and finance, is that sometimes a complete correlation matrix is not known. The reasons vary. Sometimes good estimates exist for correlations between certain pairs variables, but correlations for other pairs are unmeasured or are poor. *Any matrix that contains pairwise correlation estimates might not be a true correlation matrix* (Rousseeuw and Molenberghs 1994; Walter and Lopez 1999).

10.7.1 An Example from Finance

To give a concrete example, suppose that an analyst predicts that the correlation between certain currencies (such as the dollar, yen, and euro) will have certain values in the coming year:

- The first and second currencies will have correlation $\rho_{12} = 0.3$.

- The first and third currencies will have correlation $\rho_{13} = 0.9$.

- The second and third currencies will have correlation $\rho_{23} = 0.9$.

These estimates seem reasonable. Unfortunately, the resulting matrix of correlations is not positive definite and therefore does not represent a valid correlation matrix, as demonstrated by the following program. As shown in Figure 10.12, the matrix has a negative eigenvalue and consequently is not positive definite.

```
proc iml;
C = {1.0 0.3 0.9,
     0.3 1.0 0.9,
     0.9 0.9 1.0};
eigval = eigval(C);
print eigval;
```

Figure 10.12 Eigenvalues of an Invalid Correlation Matrix

eigval
2.4316006
0.7
-0.131601

Mathematically, the problem is that correlations between variables are not independent. They are coupled together by the requirement that a true correlation matrix is positive semidefinite. If R is a correlation matrix, then the correlations must satisfy the condition $\det(R) \geq 0$. For a 3×3 matrix, this implies that the correlation coefficients satisfy

$$\rho_{12}^2 + \rho_{13}^2 + \rho_{23}^2 - 2\rho_{12}\rho_{13}\rho_{23} \leq 1$$

The set of $(\rho_{12}, \rho_{13}, \rho_{23})$ triplets that satisfy the inequality forms a convex subset of the unit cube (Rousseeuw and Molenberghs 1994). If you substitute $\rho_{12} = 0.3$ and $\rho_{13} = \rho_{23} = 0.9$, you will discover that the inequality is not satisfied.

10.7.2   What Causes the Problem and What Can You Do?

Sometimes an estimated sample covariance or correlation matrix is not positive definite. In SAS software, this can occur when the data contain missing values. By default, the CORR procedure computes *pairwise correlations*, which means that correlations between variables are based on pairs of variables, not on the entire multivariate data. To override the default behavior, use the NOMISS option. The NOMISS option instructs PROC CORR to exclude observations for which any variable is missing, which is a process known as *listwise deletion*. This option ensures that the resulting covariance matrix is positive semidefinite.

However, if you use listwise deletion and your data contain many missing values, the covariance matrix might be a poor estimate of the population covariance. For example, if 60% of the values for one variable are missing, the covariance computation will be based on a small fraction of the total number of observations in the data.

For a similar reason, indefinite matrices also arise when estimating the polychoric correlations between ordinal variables, regardless of the presence of missing values. SAS distributes a %POLYCHOR macro that constructs a polychoric correlation matrix. (Search the `support.sas.com` Web site for Sample 25010 to obtain this macro.) The macro documentation states, "The PLCORR option in the FREQ procedure is used iteratively to compute the polychoric correlation for each pair of variables. ... The individual correlation coefficients are then assembled" into a matrix. Because the polychoric correlation is computed pairwise, the matrix might be indefinite. (Beginning with SAS 9.3, you can use PROC CORR to compute polychoric correlations; specify the POLYCHORIC option in the PROC CORR statement.)

The following options can be used to convert a matrix that is not positive definite into a matrix that is positive definite:

- One option, which is discussed in Section 10.4.3, is to increase the diagonal values of the covariance estimate by using a so-called *ridge factor* (Schafer 1997, p. 155). This results in a matrix that is diagonally dominant and therefore PD.

- A second option is to use various algorithms that are collectively known as *shrinkage* methods (Ledoit and Wolf 2004; Schäfer and Strimmer 2005). The idea is to choose a target matrix, T, which is known to be PD. (The target matrix usually has structure, such as a diagonal or compound symmetric matrix.) The sample covariance matrix, S is then "shrunk" towards the target matrix by using the linear transformation $\lambda T + (1 - \lambda)S$. This approach is not presented in this book.

- A third option is to compute the pairwise covariance matrix and, if it is not PD, find the nearest PD matrix to it (Higham 1988, 2002). This approach is popular among some researchers in finance and econometrics. This approach is presented in Section 10.8.

There are disadvantages to each method. Ridge factors and shrinkage are ad-hoc approaches that can result in a covariance matrix that is far from the data-based estimate. The Higham approach does not preserve the covariance structure of a matrix.

The next section presents an algorithm by Nick Higham (Higham 1988, 2002) that finds the closest correlation matrix (in some norm) to a given estimate. The original estimate does not, itself, need to be a valid correlation matrix.

10.8 The Nearest Correlation Matrix

The previous section demonstrated that not every symmetric matrix with unit diagonal is a correlation matrix. If you have an estimated correlation matrix that is not positive semidefinite, then you can try to project it onto the set of true correlation matrices.

Higham's method for finding the nearest correlation matrix consists of a series of alternating projections onto two sets: the set S of symmetric positive semidefinite matrices (which is actually an algebraic variety) and the set U of symmetric matrices with unit diagonals. The intersection of these two sets is the set of correlation matrices, and it is somewhat remarkable that Higham showed (using results of J. Boyle and R. Dykstra) that the alternating projections converge onto the intersection.

To project a matrix X onto the set of positive semidefinite matrices, use the spectral decomposition (see Section 8.9) to decompose the matrix as $X = QDQ`$ where Q is the matrix of eigenvectors and D is a diagonal matrix that contains the eigenvalues. Replace any negative eigenvalues with zeros and then "reassemble" the matrix, as shown in the following SAS/IML function:

```
proc iml;
/* Project symmetric X onto S={positive semidefinite matrices}.
   Replace any negative eigenvalues of X with zero */
start ProjS(X);
   call eigen(D, Q, X);              /* notice that X = Q*D*Q`     */
   V = choose(D>0, D, 0);
   W = Q#sqrt(V`);                   /* form Q*diag(V)*Q`          */
   return( W*W` );                   /* W*W` = Q*diag(V)*Q`        */
finish;
```

The projection of a matrix onto the set of matrices with unit diagonal is even easier: merely replace each of the diagonal elements with 1:

```
/* project square X onto U={matrices with unit diagonal}.
   Return X with the diagonal elements replaced by ones. */
start ProjU(X);
   n = nrow(X);
   Y = X;
   Y[do(1, n*n, n+1)] = 1;          /* set diagonal elements to 1 */
   return ( Y );
finish;
```

The main algorithm consists of calling the PROJS and PROJU functions in a loop. The simplest implementation would just call these functions 100 times and hope for the best. The following implementation keeps track of the process and stops the algorithm when the matrix has converged (within some tolerance) to a correlation matrix:

```
/* the matrix infinity norm is the max abs value of the row sums */
start MatInfNorm(A);
   return( max(abs(A[,+])) );
finish;

/* Given a symmetric matrix, A, project A onto the space of PSD
   matrices. The function uses the algorithm of Higham (2002) to
   return the matrix X that is closest to A in the Frobenius norm.  */
start NearestCorr(A);
   maxIter = 100; tol  = 1e-8;        /* initialize parameters    */
   iter = 1;       maxd = 1;          /* initial values           */
   Yold = A;  Xold = A;  dS = 0;

   do while( (iter <= maxIter) & (maxd > tol) );
     R = Yold - dS;                   /* dS is Dykstra's correction */
     X = ProjS(R);                    /* project onto S={PSD}       */
     dS = X - R;
     Y = ProjU(X);                    /* project onto U={Unit diag} */

     /* How much has X changed? (Eqn 4.1) */
     dx = MatInfNorm(X-Xold) / MatInfNorm(X);
     dy = MatInfNorm(Y-Yold) / MatInfNorm(Y);
     dxy = MatInfNorm(Y - X) / MatInfNorm(Y);
     maxd = max(dx,dy,dxy);
     iter = iter + 1;
     Xold = X;  Yold = Y;             /* update matrices            */
   end;
   return( X );                       /* X is positive semidefinite */
finish;
```

You can test the algorithm on the matrix of pairwise currency correlations. As shown in Figure 10.13, the correlation matrix has off-diagonal entries that are close to the candidate matrix, c. By slightly modifying the candidate matrix, you can obtain a true correlation matrix that can be used to simulate data with the given correlations.

```
/* finance example */
C = {1.0 0.3 0.9,
     0.3 1.0 0.9,
     0.9 0.9 1.0};
R = NearestCorr(C);
print R[format=7.4];
```

Figure 10.13 Nearest Correlation Matrix

R		
1.0000	0.3481	0.8210
0.3481	1.0000	0.8210
0.8210	0.8210	1.0000

The computational cost of Higham's algorithm is dominated by the eigenvalue decomposition. For symmetric matrices, the SAS/IML EIGEN routine is an $\mathcal{O}(n^3)$ operation, where n is the number of rows in the matrix. For small to medium-sized symmetric matrices (for example, less than 1,000 rows) the EIGEN routine in SAS/IML computes the eigenvalues in a second or two. For larger matrices (for example, 2,000 rows), the eigenvalue decomposition might take 30 seconds or more.

For large matrices (for example, more than 100 rows and columns), you might discover that the numerical computation of the eigenvalues is subject to numerical rounding errors. In other words, if you compute the numerical eigenvalues of a matrix, A, there might be some very small negative eigenvalues, such as -1×10^{-14}. If this interferes with your statistical method and SAS still complains that the matrix is not positive definite, then you can increase the eigenvalues by adding a small multiple of the identity, such as $B = \epsilon I + A$, where ϵ is a small positive value chosen so that all eigenvalues of B are positive. Of course, B is not a correlation matrix, because it does not have ones on the diagonal, so you need to convert it to a correlation matrix. It turns out that this combined operation is equivalent to dividing the off-diagonal elements by $1 + \epsilon$, as follows:

```
/* for large matrices, might need to correct for rounding errors   */
eps = 1e-10;
B = ProjU( A/(1+eps) );        /* divide off-diag elements by 1+eps */
```

Higham's algorithm is very useful. It is guaranteed to converge, although the rate of convergence is linear. A subsequent paper by Borsdorf and Higham (2010) uses a Newton method to achieve quadratic convergence.

Exercise 10.6: Use the method in Section 10.4.2 to generate a random symmetric matrix of size n. Run Higham's algorithm and time how long it takes for $n = 50, 100, 150, \dots 300$. Plot the time versus the matrix size.

10.9   References

Belsley, D. A., Kuh, E., and Welsch, R. E. (1980), *Regression Diagnostics: Identifying Influential Data and Sources of Collinearity*, New York: John Wiley & Sons.

Bendel, R. B. and Mickey, M. R. (1978), "Population Correlation Matrices for Sampling Experiments," *Communications in Statistics—Simulation and Computation*, 7, 163–182.

Borsdorf, R. and Higham, N. J. (2010), "A Preconditioned Newton Algorithm for the Nearest Correlation Matrix," *IMA Journal of Numerical Analysis*, 30, 94–107.

Davies, P. I. and Higham, N. J. (2000), "Numerically Stable Generation of Correlation Matrices and Their Factors," *BIT*, 40, 640–651.

Higham, N. J. (1988), "Computing a Nearest Symmetric Positive Semidefinite Matrix," *Linear Algebra and Its Applications*, 103, 103–118.

Higham, N. J. (1991), "Algorithm 694: A Collection of Test Matrices in MATLAB," *ACM Transactions on Mathematical Software*, 17, 289–305.
   URL http://www.netlib.org/toms/694

Higham, N. J. (2002), "Computing the Nearest Correlation Matrix—a Problem from Finance," *IMA Journal of Numerical Analysis*, 22, 329–343.

Ledoit, O. and Wolf, M. (2004), "Honey, I Shrunk the Sample Covariance Matrix," *Journal of Portfolio Management*, 30, 110–119.
URL http://ssrn.com/abstract=433840

Rapuch, G. and Roncalli, T. (2001), "GAUSS Procedures for Computing the Nearest Correlation Matrix and Simulating Correlation Matrices," Groupe de Recherche Opérationelle, Crédit Lyonnais.
URL http://thierry-roncalli.com/download/gauss-corr.pdf

Rousseeuw, P. J. and Molenberghs, G. (1994), "The Shape of Correlation Matrices," *American Statistician*, 48, 276–279.
URL http://www.jstor.org/stable/2684832

Schäfer, J. and Strimmer, K. (2005), "A Shrinkage Approach to Large-Scale Covariance Matrix Estimation and Implications for Functional Genomics," *Statistical Applications in Genetics and Molecular Biology*, 4.
URL http://uni-leipzig.de/~strimmer/lab/publications/journals/shrinkcov2005.pdf

Schafer, J. L. (1997), *Analysis of Incomplete Multivariate Data*, New York: Chapman & Hall.

Stewart, G. W. (1980), "The Efficient Generation of Random Orthogonal Matrices with an Application to Condition Estimators," *SIAM Journal on Numerical Analysis*, 17, 403–409.
URL http://www.jstor.org/stable/2156882

Walter, C. and Lopez, J. A. (1999), "The Shape of Things in a Currency Trio," Federal Reserve Bank of San Francisco Working Paper Series, Paper 1999-04.
URL http://www.frbsf.org/econrsrch/workingp/wpjl99-04a.pdf

Part IV

Applications of Simulation in Statistical Modeling

Chapter 11
Simulating Data for Basic Regression Models

Contents

11.1 Overview of Simulating Data from Basic Regression Models

This chapter describes how to simulate data for use in regression modeling. Previous chapters have described how to simulate many kinds of continuous and discrete random variables, and how to achieve correlations between those variables. All of the previous techniques can be used to generate explanatory variables in a regression model. This chapter describes several ways to construct *response variables* that are related to the explanatory variables through a regression model.

In the simplest case, the response variable is constructed as a linear combination of the explanatory variables, plus random errors. Statistical inference requires making assumptions about the random errors, such as that they are independently and normally distributed. By using the techniques in this chapter, you can explore what happens when the error terms do not satisfy these assumptions.

You can also use the techniques in this chapter for the following tasks:

- comparing the robustness and efficiency of regression techniques

- computing power curves

- studying the effect of missingness or unbalanced designs

- testing and validating an algorithm

- creating examples that demonstrate

This chapter simulates data from the following regression models:

- linear models with classification and continuous effects

- linear models with interaction effects

- regression models with outliers and high-leverage points

Chapter 12, "Simulating Data for Advanced Regression Models," describes how to simulate data from more advanced regression models.

11.2   Components of Regression Models

A regression model has three parts: the explanatory variables, a random error term, and a model for the response variable.

When you simulate a regression model, you first have to model the characteristics of the explanatory variables. For example, for a robustness study you might want to simulate explanatory variables that have extreme values. For a different study, you might want to generate variables that are highly correlated with each other.

The error term is the source of random variation in the model. When the variance of the error term is small, the response is nearly a deterministic function of the explanatory variables and the parameter estimates have small uncertainty. Use a large variance to investigate how well an algorithm can fit noisy data.

For fixed effect models, the errors are usually assumed to have mean zero, constant variance, and be uncorrelated. If you further assume that the errors are normally distributed and independent, you can obtain confidence intervals for the regression parameters. You can use simulation to compute the coverage probability of the confidence intervals when the error distribution is nonnormal. Furthermore, it is both interesting and instructive to use simulation to create heteroscedastic errors and investigate how sensitive a regression algorithm is to that assumption.

The regression model itself describes how the response variable is related to the explanatory variables and the error term. The simplest model is a response that is a linear combination of the explanatory variables and the error. More complicated models add interaction effects between variables. Even more complicated models incorporate a link function (such as a logit transformation) that relates the mean response to the explanatory variables.

11.2.1   Generating the Explanatory Variables

When simulating regression data, some scenarios require that the explanatory variables are fixed, whereas for other scenarios you can randomly generate the explanatory variables.

If you are trying to simulate data from an observational study, you might choose to model the explanatory variables. For example, suppose someone published a study that predicts the weight of a person based on their height. The published study might include a table that shows the mean and standard deviation of heights. You could assume normality and simulate data from summary statistics in the published table.

An alternative choice is to use the actual data from the original study if those data are available.

If you are trying to simulate data from a designed experiment, it makes sense to use values in the experiment that match the values in the design. For example, suppose an experiment studied the effect of nitrogen fertilizer on crop yield. If the experiment used the values 0, 25, 50, and 100 kilograms per acre, your simulation should use the same values.

11.2.2   Choosing the Variance of the Random Error Term

In a linear regression model, the response variable is assumed to be linearly related to the explanatory variable and a random normal process with mean zero and constant variance. Mathematically, the simple one-variable regression model is

$$Y_i = \beta_0 + \beta_1 X_i + \epsilon_i$$

for the observations $i = 1, 2, \ldots, n$. The error term is assumed to be normally distributed with mean zero, but what value should you use for its standard deviation?

For a simple linear regression, the root mean square error (RMSE) is an estimate of the standard deviation of the error term. If this value is known or estimated, the following statement generates the error term:

```
error = rand("Normal", 0, RMSE);  /* RMSE = std dev of error term */
```

If the original study includes categorical covariates (for example, gender of patients or treatment levels), the variance of the error term should be the same within each category although it can differ between categories.

For mixed models, the error term is usually a draw from a multivariate normal distribution with a specified covariance structure, as shown in Section 12.3.

11.3   Simple Linear Regression Models

The simplest regression model is a linear model with a single explanatory variable and with errors that are independently and normally distributed with mean zero and a constant variance. Unless otherwise specified, you can assume throughout this chapter that the errors are independent and identically distributed.

If the explanatory variable is a classification variable, the model can be analyzed by using the ANOVA procedure or GLM procedure. If the explanatory variable is continuous, the model can be analyzed by using the REG procedure or GLM procedure.

11.3.1 A Linear Model with a Single Continuous Variable

Suppose that you want to investigate the statistical properties of OLS regression. The following DATA step generates data according to the model

$$Y_i = 1 - 2X_i + \epsilon_i$$

where $\epsilon_i \sim N(0, 0.5)$.

The following DATA step simulates a sample of size 50, and then performs a regression analysis to validate that the simulated data are correct. The results are shown in Figure 11.1.

```
%let N = 50;                              /* size of each sample     */
data Reg1(keep=x y);
call streaminit(1);
do i = 1 to &N;
   x = rand("Uniform");                   /* explanatory variable    */
   eps = rand("Normal", 0, 0.5);          /* error term              */
   y = 1 - 2*x + eps;                     /* parameters are 1 and -2 */
   output;
end;
run;

proc reg data=Reg1;
   model y = x;
   ods exclude NObs;
quit;
```

Figure 11.1 Regression Analysis of Simulated Data Using the REG Procedure

The REG Procedure
Model: MODEL1
Dependent Variable: y

Analysis of Variance					
Source	DF	Sum of Squares	Mean Square	F Value	Pr > F
Model	1	15.03283	15.03283	55.90	<.0001
Error	48	12.90897	0.26894		
Corrected Total	49	27.94180			

Root MSE	0.51859	R-Square	0.5380
Dependent Mean	-0.15656	Adj R-Sq	0.5284
Coeff Var	-331.24481		

Parameter Estimates					
Variable	DF	Parameter Estimate	Standard Error	t Value	Pr > |t|
Intercept	1	0.94547	0.16464	5.74	<.0001
x	1	-1.94566	0.26024	-7.48	<.0001

Figure 11.1 indicates that the simulated response, y, was generated according to the regression model. The RMSE is close to 0.5, which is the standard deviation of the random error term, and the parameter estimates are close to the population values of 1 and -2.

Notice that the x variable is randomly generated for this simulation. You could also have used uniformly spaced points (`x=i`) or normally distributed values, depending on the scenario that you are simulating.

How you generate the x variable depends on what you are trying to do:

- If you intend for x to be a *fixed effect*, the values of x should be generated (or read from a data set) one time at the beginning of the simulation. For each sample, the model for the response is computed by using the same values of x.

- If you think of x as a *random effect*, then the values of x should be simulated for each sample in the simulation.

If you are just generating some "quick and dirty" data to use to test some algorithm, then it might not matter which option you choose.

The next sections assume that you want to simulate a fixed effect. Consequently, you should generate values for the explanatory variable one time and retain those values during the simulation.

11.3.1.1   Using Arrays to Hold Explanatory Variables

One way to simulate a fixed effect is to use an array of length N. For each simulated sample, the same values of **x** are used to compute **y**.

```
/* Simulate multiple samples from a regression model */
/* Technique 1: Put explanatory variables into arrays */
%let N = 50;                        /* size of each sample     */
%let NumSamples = 100;              /* number of samples       */
data RegSim1(keep= SampleID x y);
array xx{&N} _temporary_;          /* do not output the array */
call streaminit(1);
do i = 1 to &N;                    /* create x values one time */
   xx{i} = rand("Uniform");
end;

do SampleID = 1 to &NumSamples;
   do i = 1 to &N;
      x = xx{i};                   /* use same values for each sample */
      y = 1 - 2*x + rand("Normal", 0, 0.5); /* params are 1 and -2 */
      output;
   end;
end;
run;
```

This technique is also useful for simulating time series (see Chapter 13) because you can easily formulate models that use lagged values of the x variable. A minor drawback of this technique is that you need an array to store each explanatory variable. If you are generating a large number of variables, you might find it convenient to use a two-dimensional DATA step array.

11.3.1.2 Changing the Order of Loops

An alternative way to simulate a fixed effect is to reverse the usual order of the DO loops. The following example has an outer loop over observations and an inner loop for each sample:

```
/* Technique 2: Put simulation loop inside loop over observations */
data RegSim2(keep= SampleID i x y);
call streaminit(1);

do i = 1 to &N;
   x = rand("Uniform");           /* use this value NumSamples times */
   eta = 1 - 2*x;                 /* parameters are 1 and -2         */
   do SampleID = 1 to &NumSamples;
      y = eta + rand("Normal", 0, 0.5);
      output;
   end;
end;
run;


proc sort data=RegSim2;
   by SampleID i;
run;
```

For this method, you must sort by the SampleID variable prior to running a BY-group analysis. Sorting by the observation number (the i variable) is not always necessary, but it is helpful to know that each simulated sample contains the observations in the same order. For time series data, sorting by both variables is required.

No matter which method you use to simulate the data, you can use the techniques that are described in Chapter 5, "Using Simulation to Evaluate Statistical Techniques," to analyze the data. The following analysis examines the distributions of the parameter estimates:

```
proc reg data=RegSim2 outest=OutEst NOPRINT;
   by SampleID;
   model y = x;
quit;


ods graphics on;
proc corr data=OutEst noprob plots=scatter(alpha=.05 .1 .25 .50);
   label x="Estimated Coefficient of x"
         Intercept="Estimated Intercept";
   var Intercept x;
   ods exclude VarInformation;
run;
```

Figure 11.2 summarizes the approximate sampling distribution (ASD) of each parameter estimate. The output also shows that the estimates are strongly correlated with each other. The sample means for the parameter estimates (the Monte Carlo estimates) are extremely close to $(1, -2)$, which are the population parameters. The Pearson correlation for the parameter estimates is about -0.89. These two facts are shown graphically in Figure 11.3, which shows a scatter plot of the parameter estimates along with probability contours for a bivariate normal distribution.

Figure 11.2 Approximate Sampling Distribution of Parameter Estimates

The CORR Procedure

Simple Statistics							
Variable	N	Mean	Std Dev	Sum	Minimum	Maximum	Label
Intercept	100	0.99691	0.14611	99.69056	0.68939	1.43054	Estimated Intercept
x	100	-1.98239	0.27972	-198.23949	-2.70077	-1.25532	Estimated Coefficient of x

Pearson Correlation Coefficients, N = 100		
	Intercept	x
Intercept Estimated Intercept	1.00000	-0.88963
x Estimated Coefficient of x	-0.88963	1.00000

Figure 11.3 ASD of Parameter Estimates, $N = 100$

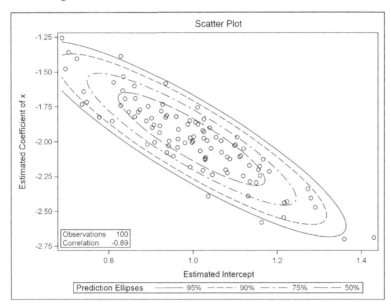

Exercise 11.1: Compare the computational efficiency of the "array method" (Section 11.3.1.1) and the "reverse loops and sort method." Use 10,000 samples. Which technique is faster? Why?

Exercise 11.2: The OutEst data set also contains estimates for the RMSE for each simulated set of data. Use PROC UNIVARIATE to analyze the _RMSE_ variable. What is the Monte Carlo estimate for the RMSE? What is a 90% confidence interval? Test whether the ASD is normally distributed.

11.3.2 A Linear Model Based on Real Data

The previous section generated random values for the explanatory variable. This section describes ways to incorporate a model of real data into a simulation.

The "Getting Started" section for the SAS documentation on the REG procedure in the *SAS/STAT User's Guide* describes a regression analysis of weight as a function of height for 19 students. The following SAS statements reproduce two tables in the analysis:

```
ods graphics off;
proc reg data=Sashelp.Class;
   model Weight = Height;
   ods select FitStatistics ParameterEstimates;
quit;
```

Figure 11.4 Fit Statistics and Parameter Estimates for a Linear Model

**The REG Procedure
Model: MODEL1
Dependent Variable: Weight**

Root MSE	11.22625	R-Square	0.7705
Dependent Mean	100.02632	Adj R-Sq	0.7570
Coeff Var	11.22330		

Parameter Estimates					
Variable	DF	Parameter Estimate	Standard Error	t Value	Pr > |t|
Intercept	1	-143.02692	32.27459	-4.43	0.0004
Height	1	3.89903	0.51609	7.55	<.0001

If you assume that the response variable, Weight, can be predicted by a linear function of Height, then the regression model that best fits the data is approximately

$$\text{Weight} = -143 + 3.9 \times \text{Height}$$

The RMSE for the model is 11.23, so that value can be used for the standard deviation of an error term: $\epsilon \sim N(0, 11.23)$. The following sections describe several ways to simulate these data.

11.3.2.1 Using the Sample Data

The first approach uses the exact student heights for the simulation, but uses the regression model to simulate new responses (weights). You can imagine the simulated sample to be a matched cohort: You find 19 "new students" in the population that have exactly the same heights as the original students, and you "measure" their weights. The following DATA step reads in the original heights and simulates the weights:

```
data StudentData(keep= Height Weight);
call streaminit(1);
set Sashelp.Class;            /* implicit loop over observations    */
b0 = -143; b1 = 3.9;          /* parameter estimates from regression */
rmse = 11.23;                 /* estimate of scale of error term     */
Weight = b0 + b1*Height + rand("Normal", 0, rmse);
run;
```

Notice that the linear expression `b0 + b1*Height` computes the predicted values of the regression. Consequently, an alternative but equivalent approach is to output the predicted values from PROC REG when you run the analysis. You can then generate new responses by adding a random term to the predicted value. An advantage to this approach is that you do not need to specify the regression coefficients explicitly.

Exercise 11.3: Use the P= option in the OUTPUT statement of the REG procedure to output the predicted values. Simulate new responses by adding a random error to each predicted value.

11.3.2.2   Simulations That Use the Sample Data

To generate multiple samples from the student data, add a DO loop to the programs in the previous section. The following DATA step reads the original Sashelp.Class data and simulates **NumSamples** values of the Weight variable for each observation in the data set. The Weight variable is simulated by using the regression model for the observed data.

```
/* duplicate data by using sequential access followed by a sort    */
%let NumSamples = 100;              /* number of samples            */
data StudentSim(drop= b0 b1 rmse);
b0 = -143; b1 = 3.9; rmse = 11.23;  /* parameter estimates          */
call streaminit(1);
set Sashelp.Class;                 /* implicit loop over obs       */
i = _N_;
eta = b0 + b1*Height;              /* linear predictor for student */
do SampleID = 1 to &NumSamples;
   Weight = eta + rand("Normal", 0, rmse);
   output;
end;
run;

proc sort data=StudentSim;
   by SampleID i;
run;
```

You can avoid the inverted loops and the subsequent sorting if you use the POINT= option in the SET statement to randomly access observations. If you use this approach, then be sure to use the STOP statement after both loops have completed in order to prevent an infinite loop. The DATA step does not detect an end-of-file condition when you use random access, so you must explicitly stop the processing of data.

```
/* Alternative: simulate weights by using random access and no sort */
data StudentSim2(drop= b0 b1 rmse);
b0 = -143; b1 = 3.9; rmse = 11.23;   /* parameter estimates          */
call streaminit(1);
do SampleID = 1 to &NumSamples;
   do i = 1 to NObs;                    /* NObs defined at compile time */
      set Sashelp.Class point=i nobs=NObs;      /* random access */
      eta = b0 + b1*Height;          /* linear predictor for student */
      Weight = eta + rand("Normal", 0, rmse);
      output;
   end;
end;
STOP;     /* IMPORTANT: Use STOP with POINT= option in the SET stmt */
run;
```

Exercise 11.4: Call PROC REG on the StudentSim data, and use SampleID as a BY-group variable in order to generate the ASD of the regression coefficients.

Exercise 11.5: The second DATA step in this section reads the Sashelp.Class multiple times. Use the technique in Section 11.3.1.1 to read the Height variable into an array and access the array when building the linear model. Use a macro variable to hold the number of observations in the data set.

Exercise 11.6: The Sashelp.Cars data set has 428 observations. Generate 10,000 copies of the data by using each of the two techniques in this section. Compare the performance of the two methods. Which is faster?

11.3.2.3   Simulating Data from Regression Models in SAS/IML Software

You can use the SAS/IML language to simulate data from regression models. Because the SAS/IML language supports matrix and vector computations, the model is represented by the matrix equation $Y = X\beta + \epsilon$ where X is an $N \times p$ design matrix, β is a p-dimensional vector of regression parameters, and ϵ is an N-dimensional random vector such that each element is independently chosen from $N(0, \sigma)$.

In order to include an intercept term in the model, the first column of X is a column of 1s. The model $Y_i = \beta_0 + \beta_1 X_i + \epsilon_i$ for $i = 1, \ldots, N$ can be rewritten in vector notation as $Y_i = \begin{bmatrix} 1 & X_i \end{bmatrix} \beta + \epsilon_i$ where $\beta = \begin{pmatrix} \beta_0 & \beta_1 \end{pmatrix}'$.

The following SAS/IML program defines $\beta = \begin{pmatrix} -143 & 3.9 \end{pmatrix}'$ and forms the linear predictor:

```
proc iml;
call randseed(1);
beta = {-143, 3.9};   rmse = 11.23;

use Sashelp.Class NOBS N;             /* N = sample size        */
read all var {Height} into X1;        /* read data              */
close Sashelp.Class;

X = j(N,1,1) || X1;                   /* add intercept column   */
eta = X*beta;                         /* linear predictor       */
```

There are several ways to simulate the response variable. If you are going to analyze the simulated data in the SAS/IML language, then you can just loop over the number of simulations, as follows:

```
eps = j(N, 1);                            /* allocate N x 1 vector   */
do i = 1 to &NumSamples;
   call randgen(eps, "Normal", 0, rmse); /* fill with random normal */
   Weight = eta + eps;                    /* one simulated response  */
   /* conduct further analysis */
end;
```

On the other hand, if you intend to write the data to a SAS data set for analysis by one of the regression procedures, you might want to generate all of the simulated data in "long" format, as follows:

```
eps = j(N * &NumSamples, 1);              /* allocate long vector    */
call randgen(eps, "Normal", 0, rmse);     /* fill with random normal */
Wt = repeat(eta, &NumSamples) + eps;      /* simulate all responses  */
ID = repeat( T(1:&NumSamples), 1, N );    /* 1,1,1,...,2,2,2,...      */
Ht = repeat( X1, &NumSamples );
create SimReg var {ID Ht Wt};  append;  close SimReg; /* write data */
```

Exercise 11.7: Use PROC REG and a BY statement to analyze the simulated data. Compute the ASD for the parameter estimates.

11.3.2.4 Modeling the Data

Suppose that you do not have access to the original data, but that you are told that Figure 11.4 is the result of a regression on 19 students with mean height of 62.3 inches and standard deviation of 5.13 inches. If you assume that the heights are normally distributed, the following example creates simulated data with 19 observations. The heights and weights are not the same as for the original study, but are for a new set of 19 simulated students.

```
/* Original data not available. Simulate from summary statistics.  */
data StudentModel(keep= Height Weight);
call streaminit(1);
b0 = -143; b1 = 3.9;         /* parameter estimates from regression */
rmse = 11.23;                /* estimate of scale of error term    */
do i = 1 to 19;
   Height = rand("Normal", 62.3, 5.13);  /* Height is random normal */
   Weight = b0 + b1*Height + rand("Normal", 0, rmse);
   output;
end;
run;
```

The parameter estimates from the regression analysis are used to simulate a new data set with similar characteristics. The distribution of heights is modeled by a normal distribution.

This example shows a common technique in simulation: using parameter estimates as parameters. This approach is not perfect—for example, the Monte Carlo estimates are centered on the estimates instead of on the parameters—but because the parameter values are unknown, this is often the best that you can achieve.

An advantage to this approach is that you can use it to generate samples that are a different size than the original data sample. For example, you can change the upper bound for the DO loop from 19 to 50 to simulate heights and weights in a class that contains 50 students.

Exercise 11.8: Use the modeling approach to generate 1,000 samples of student heights and weights. For each sample, generate 19 "new" students with heights that are normally distributed. Analyze the distribution of the parameter estimates.

11.3.3   A Linear Model with a Single Classification Variable

The SAS documentation on the ANOVA procedure in the *SAS/STAT User's Guide* shows a "Getting Started" example of a one-way analysis of variance for a balanced design. In the example, the response variable depends on a classification variable that has six levels. There are five observations for each level. The assumed model is

$$Y_{ij} = \mu + \alpha_i + \epsilon_{ij}$$

where $i = 1, \ldots, 6$ enumerates the treatment levels, $\epsilon_{ij} \sim N(0, \sigma_j)$, and $j = 1, \ldots, 5$ enumerates the observations in each treatment group.

Assume that the overall mean is 20 and that the effect of the six treatment levels are 9, -6, -6, 4, 0, and 0. Assume that the corresponding standard deviations are 6, 2, 4, 4, 1, and 2. (These values are loosely based on an experiment that is described in the PROC ANOVA documentation.) The following DATA step simulates data from this balanced design:

```
data AnovaData(keep=Treatment Y);
call streaminit(1);
grandmean = 20;
array effect{6} _temporary_ (9 -6 -6 4 0 0);
array std{6}    _temporary_ (6  2  4 4 1 2);
do i = 1 to dim(effect);      /* number of treatment levels       */
   Treatment = i;
   do j = 1 to 5;             /* number of obs per treatment level */
      Y = grandmean + effect{i} + rand("Normal", 0, std{i});
      output;
   end;
end;
run;
```

The AnovaData data set contains simulated data from the regression model. You can call PROC ANOVA to analyze the simulated data, as follows. Figure 11.5 shows the results.

```
proc ANOVA data=AnovaData;
   class Treatment;
   model Y = Treatment;
   ods exclude ClassLevels NObs;
run;
```

Figure 11.5 ANOVA Analysis of Simulated Data

The ANOVA Procedure

Dependent Variable: Y

Source	DF	Sum of Squares	Mean Square	F Value	Pr > F
Model	5	1111.478287	222.295657	38.29	<.0001
Error	24	139.340596	5.805858		
Corrected Total	29	1250.818883			

R-Square	Coeff Var	Root MSE	Y Mean
0.888601	11.28273	2.409535	21.35596

Source	DF	Anova SS	Mean Square	F Value	Pr > F
Treatment	5	1111.478287	222.295657	38.29	<.0001

Each statistic in Figure 11.5 has a sampling distribution. Recall from Chapter 5 that you can simulate a sampling distribution by adding an additional DO loop to the DATA statement, such as

```
do SampleID = 1 to &NumSamples; /* simulation loop */
   ...
end;
```

and by using a **BY SampleID** statement in the call to PROC ANOVA.

You can also simulate the regression data in SAS/IML software. Whereas the typical DATA step technique is to generate two columns of data (Treatment and Y), the preferred technique in the SAS/IML language is to construct a 6×5 matrix where each row represents a sample for a treatment level, as shown in the following program:

```
proc iml;
call randseed(1);
grandmean = 20;
effect = {9 -6 -6 4 0 0};
std    = {6  2  4 4 1 2};
N = 5;

/* each row of Y is a treatment */
Y = j(ncol(effect),N,.);   /* allocate matrix; fill with missing values */
ei = j(1,N);

do i = 1 to ncol(effect);
   call randgen(ei, "Normal", 0, std[i]);
   Y[i,] = grandmean + effect[i] + ei;
end;
```

Exercise 11.9: Choose a statistic in the FitStatistics table in Figure 11.5, such as the R-square statistic, the coefficient of variation, or the RMSE. Use PROC ANOVA and the techniques from Chapter 5 to investigate the ASD for these statistics.

Exercise 11.10: In an unbalanced design, the number of observations is not constant among treatment groups. Suppose that the experiment uses sample sizes of 5, 10, 10, 12, 6, and 8 for the treatment groups.

1. Write a DATA set that simulates data with these properties.

2. Modify the SAS/IML program to simulate these data. One approach is to construct a $6 \times K$ matrix, where K is the maximum sample size in the treatment groups, and use missing values to represent elements that are not part of the design.

11.3.4   A Linear Model with Classification and Continuous Variables

Sometimes you need a quick and easy way to simulate data from a regression model with many classification variables, many continuous variables, and a response variable. Suppose that you want to generate a predetermined number of continuous and classification variables, and you want each classification variable to have L levels. The following DATA step simulates (uncorrelated) data with these properties:

```
%let nCont = 4;                   /* number of contin vars           */
%let nClas = 2;                   /* number of class vars            */
%let nLevels = 3;                 /* number of levels for each class var */
%let N = 100;
/* Simulate GLM data with continuous and class variables */
data GLMData(drop=i j);
array x{&nCont} x1-x&nCont;
array c{&nClas} c1-c&nClas;
call streaminit(1);

/* simulate the model */
do i = 1 to &N;
   do j = 1 to &nCont;            /* continuous vars for i_th obs */
      x{j} = rand("Uniform");     /* uncorrelated uniform        */
   end;
   do j = 1 to &nClas;            /* class vars for i_th obs      */
      c{j} = ceil(&nLevels*rand("Uniform"));   /* discrete uniform */
   end;
   /* specify regression coefficients and magnitude of error term */
   y = 2*x{1} - 3*x{&nCont} + c{1} + rand("Normal");
   output;
end;
run;
```

In this model, all continuous variables are uniformly distributed. Furthermore, the discrete levels of the classification variables are also uniformly distributed. The response variable in the example does not depend on all of the explanatory variables. It depends only on the first and last continuous variables and on the first classification variable. Consequently, if you analyze the simulated model, then there should be only three regression coefficients that are significant. The following statements use PROC GLM to analyze the data. The Type III sums of squares are shown in Figure 11.6. The three variables that were used to construct the response variable have p-values that are less than 0.0001.

```
proc glm data=GLMData;
   class c1-c&nClas;
   model y = x1-x&nCont c1-c&nClas / SS3;
   ods select ModelANOVA;
quit;
```

Figure 11.6 GLM Analysis of Simulated Data

The GLM Procedure

Dependent Variable: y

Source	DF	Type III SS	Mean Square	F Value	Pr > F
x1	1	16.83531580	16.83531580	18.92	<.0001
x2	1	0.00466132	0.00466132	0.01	0.9425
x3	1	0.00001342	0.00001342	0.00	0.9969
x4	1	62.17137325	62.17137325	69.87	<.0001
c1	2	69.89843326	34.94921663	39.28	<.0001
c2	2	0.92738148	0.46369074	0.52	0.5956

Exercise 11.11: In the simulated data, each classification variable contains three levels. Modify the DATA step to simulate three classification variables for which the number of levels are 2, 3, and 4, respectively.

11.4   The Power of a Regression Test

This section presents a case study that shows how to use simulation to investigate the power of a statistical test.

When examining regression models, it is common to ask whether the coefficient of an explanatory effect is significantly different from zero. The TEST statement in the REG procedure can be used to test whether a regression coefficient is zero, given the data.

To demonstrate the TEST statement, the following statements simulate a single sample of data for a regression model for which the coefficient of the z variable is exactly zero. The results are shown in Figure 11.7.

```
data Reg1(drop=i);
call streaminit(1);
do i = 1 to &N;
   x = rand("Normal");
   z = rand("Normal");
   y = 1 + 1*x + 0*z + rand("Normal");        /* eps ~ N(0,1) */
   output;
end;
run;

proc reg data=Reg1;
   model y = x z;
   z0: test z=0;
   ods select TestANOVA;
quit;
```

Figure 11.7 Test whether a Regression Parameter Is Zero

The REG Procedure
Model: MODEL1

Test z0 Results for Dependent Variable y				
Source	DF	Mean Square	F Value	Pr > F
Numerator	1	0.06601	0.07	0.7982
Denominator	97	1.00417		

The TestANOVA table, which is shown in Figure 11.7, shows the result of an F test for the null hypothesis that the coefficient of the z variable is zero. The p-value for the F test is large, which indicates that there is little evidence to reject the null hypothesis. This result is expected, because the data were simulated so that the null hypothesis is true.

You can conduct a simulation that examines the power of the TEST statement to detect small (but nonzero) values of the regression coefficient. The F test assumes that the error term of the regression model is normally distributed and is homoscedastic. But how does the F test behave if these assumptions are violated?

A simulation that is presented in Greene (2000, Ch. 15) examines this question. Greene examines three different error distributions for the error term in the regression model:

- The error term is normally distributed. This is the usual assumption that, if true, implies that the statistic in the TestANOVA table follows an F distribution.

- The error term follows a t distribution with 5 degrees of freedom. This distribution is similar to a normal distribution, but has fatter tails.

- The error term is heteroscedastic. Greene (2000) uses a normal distribution in which the standard deviation is $\exp(0.2x)$, which depends on the values of the x variable. Greene (2000, p. 617) comments that "the statistic is entirely wrong if the disturbances are heteroscedastic."

The design of Greene's simulation is to draw 50 uncorrelated observations of two variables, x and z, from $N(0, 1)$. For each observation, compute the true model as $\eta_i = 1 + x_i + \gamma z_i$ and compute $y_i = \eta_i + \epsilon_i$ for each of the three kinds of error distributions. Repeat this simulation for a sequence of γ values in the range $[0, 1]$. The following DATA step simulates data according to this scheme:

```
%let N = 50;
%let NumSamples = 1000;                          /* number of samples    */
data RegSim(drop= i eta sx);
call streaminit(1);
do i = 1 to &N;
   x = rand("Normal");   z = rand("Normal");
   sx = exp(x/5);                                /* StdDev for model 3   */
   do gamma = 0 to 1 by 0.1;
      eta = 1 + 1*x + gamma*z;                   /* linear predictor     */
      do SampleID = 1 to &NumSamples;
         /* Model 1: e ~ N(0,1)        */
         /* Model 2: e ~ t(5)          */
         /* Model 3: e ~ N(0, exp(x/5)) */
         Type = 1;  y = eta + rand("Normal");          output;
         Type = 2;  y = eta + rand("T", 5);            output;
         Type = 3;  y = eta + rand("Normal", 0, sx);   output;
      end;
   end;                                          /* end gamma loop       */
end;                                             /* end observation loop */
run;

proc sort data=RegSim out=Sim;
   by Type gamma SampleID;
run;
```

As described in Section 11.3.1.2, the order of the DO loops in this simulation requires that you sort the data prior to running the regression analysis.

The following PROC REG statement analyzes each set of simulated data for each kind of error distribution (Type), for each value of the regression coefficient (gamma), and for each of 1,000 samples (SampleID):

```
/* Turn off output when calling PROC for simulation */
%ODSOff
proc reg data=Sim;
   by Type gamma SampleID;
   model y = x z;
   test z=0;
   ods output TestANOVA=TestAnova;
quit;
%ODSOn
```

The following statements count the number of times that the null hypothesis ($\gamma = 0$) was rejected for each kind of error distribution for each value of γ:

```
/* 3. Construct an indicator variable for observations that reject H0 */
data Results;
   set TestANOVA(where=(Source="Numerator"));
   Reject = (ProbF <= 0.05);                      /* indicator variable */
run;

/* count number of times H0 was rejected */
proc freq data=Results noprint;
   by Type gamma;
   tables Reject / nocum out=Signif(where=(reject=1));
run;

data Signif;
   set Signif;
   proportion = percent / 100;      /* convert percent to proportion */
run;
```

You can plot the proportion of times that the null hypothesis is rejected. For $\gamma > 0$, the test commits a Type II error if it fails to reject the null hypothesis. The results are shown in Figure 11.8:

```
proc format;
   value ErrType 1="e ~ N(0,1)"  2="e ~ t(5)"  3="e ~ Hetero";
run;

title "Power of F Test for gamma=0";
title2 "N = &N";
proc sgplot data=Signif;
   format Type ErrType.;
   series x=gamma y=proportion / group=Type;
   yaxis min=0 max=1 label="Power (1 - P[Type II Error])" grid;
   xaxis label="Gamma" grid;
   keylegend / across=1 location=inside position=topleft;
run;
```

Figure 11.8 Power Curves for Testing whether a Coefficient Is Zero

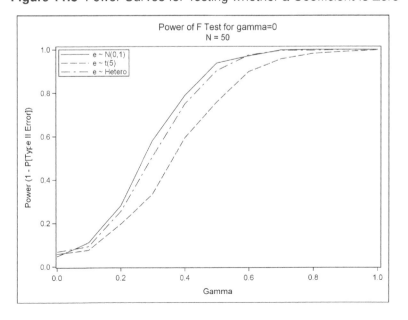

Figure 11.8 shows how the power curve for the F test varies for the three error distributions. Greene concludes that "it appears that the presence of heteroscedasticity [does not] degrade the power of the statistic. But the different distributional assumption does."

Exercise 11.12: Use DATA step arrays, as described in Section 11.3.1.1, to rewrite the DATA step so that no sort is required.

Exercise 11.13: Set $\gamma = 0$ in the regression model. When the error term follows an N(0,1) distribution, the F statistic in the TestANOVA table follows an $F_{2,N-2}$ distribution. Examine the sampling distribution for the other choices of the error distribution. Does the F statistic appear to be sensitive to the shape of the error distribution?

11.5   Linear Models with Interaction and Polynomial Effects

Previous sections have simulated data from regression models with main effects. This section shows how to simulate data with interaction and polynomial effects.

11.5.1   Polynomial Effects

It is easy to simulate data that contain a polynomial effect. For example, for the regression model

$$Y_i = 1 - 2X_i + 3X_i^2 + \epsilon_i$$

you can use the following DATA step syntax:

```
y = 1 - 2*x + 3*x*x + eps;            /* include quadratic effect */
```

For multivariate polynomials, you can create interactions between two continuous variables simply by multiplying the variables together. In the SAS/IML language, use the elementwise multiplication operator (#) to multiply the elements of two vectors.

Exercise 11.14: Simulate 1,000 observations from the two-variable regression model $Y_i = 1 - 2X_i + 3Z_i + X_i^2 - X_i Z_i + \epsilon_i$, where $\epsilon \sim N(0, 1)$. Run the GLM procedure to verify that the simulated data are from the specified model.

11.5.2   Design Matrices

Many interesting models include interactions between variables, and it is not always easy to use the DATA step to specify the interactions. An alternative approach is to use SAS/STAT procedures to create a design matrix for the effects. A design matrix is a unified way to incorporate classification variables, continuous variables, and interactions. You can then read the design matrix into PROC IML and simulate the data by using the matrix model $Y = X\beta + \epsilon$. Here X is an $N \times p$ design matrix, β is a p-dimensional vector of regression parameters, and ϵ is an N-dimensional random vector with zero mean.

There are various parameterizations that you can use to create dummy variables from classification variables. The most common parameterization is the GLM parameterization (also called a *singular parameterization*), which you can generate by using the GLMMOD procedure. The LOGISTIC procedure supports alternate parameterizations. You can create design matrices for models that include continuous variables, but the examples in this section only include classification variables and their interactions.

11.5.2.1   Design Matrices with GLM Parameterization

Suppose that you want to simulate data for a study that has two classification variables. The Drug variable contains three levels and the Disease variable contains two levels. Furthermore, you want the mean response to depend on the explanatory variables according to Table 11.1.

Table 11.1   Mean Response for Joint Levels

Drug	Disease 1	Disease 2
1	10	0
2	15	-5
3	20	-10

The following DATA step creates a balanced design matrix with five repeated measurements for each joint level of Drug and Disease. The value of the response variable (y) is arbitrary and unimportant, because the sole purpose of this DATA step is to generate a design matrix for the explanatory variables.

```
data Interactions;
y = 0;                             /* the value of y does not matter */
do drug = 1 to 3;
   do disease = 1 to 2;
      do subject = 1 to 5;
         output;
      end;
   end;
end;
run;
```

You can use the GLMMOD procedure to generate a design matrix from the Interactions data set. The GLMMOD procedure enables you to specify the effects (including interaction terms), and it generates a design matrix that corresponds to the model. For example, the following statements generate a design matrix for a model for which the response variable depends on Drug, Disease, and their interaction:

```
proc glmmod data=Interactions noprint
         outparm=Parm outdesign=Design(drop=y);        /* DROP y */
   class drug disease;
   model y = drug | disease;
run;
```

The GLMMOD procedure creates two data sets. The Parm data set maps the 12 columns of the design matrix to the corresponding variable names and levels. The Design data set contains the design matrix. Notice that the DROP= data set option is used to prevent the y variable from appearing in the Design data. The columns of the Design data set are named Col1–Col12.

You can construct a parameter vector of regression coefficients and use matrix multiplication in PROC IML to form the linear model and to add a random error term, as shown in the following statements:

```
proc iml;
call randseed(1);
use Design;
read all var _NUM_ into X;
close Design;

/* Intcpt |--Drug--|Disease|-----Interactions-----| */
beta = {0, 0, 0, 0,  0, 0,  10, 0, 15, -5, 20, -10};
eps = j(nrow(X),1);                     /* allocate error vector */
call randgen(eps, "Normal");
y = X*beta + eps;

create Y var{y};  append;  close Y;        /* write to data set    */
```

In this example, the design matrix is rank 6, which means that there are six linearly independent columns. Consequently, there are many equivalent ways to specify the linear model that relates the design matrix to y. The approach used in the program makes it clear that Table 11.1 describes the expected value of the response for each joint level of the Drug and Disease variables.

To check that the simulated responses are correct, run the GLM procedure and graph the observed and predicted values for each level of the classification variables, as shown in the following statements:

```
data D;
   merge Y Interactions(drop=y);
run;

ods graphics on;
proc glm data=D;
   class drug disease;
   model y = drug | disease / solution p;
run;
```

Figure 11.9 shows that the predicted values for Disease=1 are approximately 10, 15, and 20. The predicted values for Disease=2 are approximately 0, −5, and −10. This shows close agreement with Table 11.1.

Because the design matrix is singular, the parameter estimates found by PROC GLM might not be the same as the parameter values that were used to construct the data. However, the predicted values (which are not shown) are independent of the parameterization.

Figure 11.9 Graph of Observed and Predicted Values

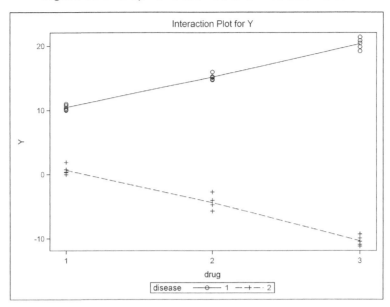

11.5.2.2   Design Matrices for Alternative Parameterizations

The GLMMOD procedure creates a singular design matrix, which means that the parameter estimates are not unique. Several SAS procedures support nonsingular parameterizations for classification effects. Two popular alternative parameterizations are the "effect" and "reference" parameterizations. You can generate design matrices for these and other parameterizations by using the LOGISTIC procedure.

The PROC LOGISTIC statement supports the OUTDESIGN= and OUTDESIGNONLY options. You can use these options to generate a design matrix without actually fitting a logistic model. Because no model is fit, you can set the values of the response variable to be a constant value, such as 0. Notice that the Interactions data set (which was created in Section 11.5.2.1) has a constant value for Y. Consequently, the following PROC LOGISTIC statement generates a design matrix by using the reference parameterization. The reference parameterization is specified by the PARAM=REFERENCE option in the CLASS statement.

```
proc logistic data=Interactions
              outdesignonly outdesign=DesignRef(drop=y);
   class drug disease / param=reference;
   model y = drug | disease;
run;
```

The PROC LOGISTIC statements write the design matrix to the DesignRef data set. The design matrix has only six columns, whereas the design matrix that is created by PROC GLMMOD has 12 columns. For details about nonsingular parameterizations and how to specify them, see the section "Parameterization of Model Effects" in the chapter "Shared Concepts and Topics" in the *SAS/STAT User's Guide*.

11.6   Outliers and Robust Regression Models

A common simulation task is to assess the robustness of statistical methods in the presence of outliers. In a regression model, the word *outlier* means an outlier for the response variable. This is an observation for which the observed response is much different than would be predicted by a robust fit of the data.

Observations can also have highly unusual values for the explanatory variables. These observations are called *high-leverage points* because they can unduly influence parameter estimates for an ordinary least squares fit.

This section describes how to simulate regression data with outliers and high-leverage points.

11.6.1   Simulating Outliers

This section describes how to generate extreme values (outliers) for a response variable. For simplicity, suppose that you want to simulate data for a univariate regression model that is given by

$$Y_i = 1 - 2X_i + \epsilon_i$$

where ϵ_i follows a contaminated normal distribution as described in Section 7.5.2. (In general, you can choose any long-tailed distribution for ϵ.) For concreteness, assume that ϵ_i follows an $N(0, 1)$ distribution 90% of the time and an $N(0, 10)$ distribution 10% of the time. The following DATA step simulates one sample:

```
%let N = 100;                            /* size of each sample     */
data RegOutliers(keep=x y Contaminated);
array xx{&N} _temporary_;
p = 0.1;                                 /* prob of contamination   */
call streaminit(1);
/* simulate fixed effects */
do i = 1 to &N;
   xx{i} = rand("Uniform");
end;
/* simulate regression model */
do i = 1 to &N;
   x = xx{i};
   Contaminated = rand("Bernoulli",p);
   if Contaminated then eps = rand("Normal", 0, 10);
   else               eps = rand("Normal", 0, 1);
   y = 1 - 2*x + eps;                    /* parameters are 1 and -2  */
   output;
end;
run;
```

You can use PROC SGPLOT to plot the data and the model. Figure 11.10 shows that most of the simulated data fall close to the line $y = 1 - 2x$. The line is displayed by using the LINEPARM statement. Nine points are outliers.

```
proc format;
   value Contam 0="N(0, 1)"  1="N(0, 10)";
run;

proc sgplot data=RegOutliers(rename=(Contaminated=Distribution));
   format Distribution Contam.;
   scatter x=x y=y / group=Distribution;
   lineparm x=0 y=1 slope=-2;                    /* requires SAS 9.3 */
run;
```

Figure 11.10 Graph of Simulated Data with 10% Contaminated Errors

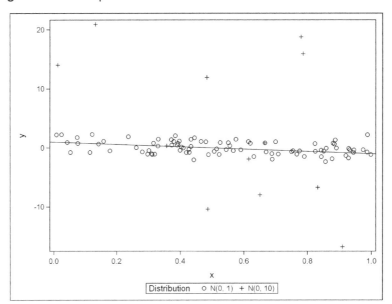

If you exclude the points for which Contaminated=0, then the parameter estimates are close to the parameter values. Of course, in practice you do not know which observations are contaminated. However, you can use the ROBUSTREG procedure to produce a robust regression estimate of the parameters. The following statements use the least trimmed squares (LTS) method to identify outliers. The FWLS option displays the final weighted least squares (FWLS) estimates (shown in Figure 11.11), which are the OLS estimates after the outliers are set to have zero weight.

```
proc robustreg data=RegOutliers method=lts FWLS;
   model y = x;
   ods select LTSEstimates ParameterEstimatesF;
run;
```

For these simulated data, both the LTS estimates and the FWLS estimates are somewhat close to the parameter values. The two estimates are usually similar, but are computed in different ways. The LTS estimates are based on a subset of the data (by default, about 75%). The LTS estimates are used to determine which observations are outliers. The FWLS estimates are ordinary least squares estimates that exclude the outliers.

Figure 11.11 Robust Parameter Estimates

The ROBUSTREG Procedure

LTS Parameter Estimates		
Parameter	DF	Estimate
Intercept	1	1.5669
x	1	-2.4405
Scale (sLTS)	0	1.2661
Scale (Wscale)	0	1.0946

Parameter Estimates for Final Weighted Least Squares Fit							
Parameter	DF	Estimate	Standard Error	95% Confidence Limits		Chi-Square	Pr > ChiSq
Intercept	1	0.8492	0.2451	0.3688	1.3297	12.00	0.0005
x	1	-1.5470	0.4039	-2.3386	-0.7553	14.67	0.0001
Scale	0	1.0409					

By using the technique in this section, you can simulate data that contain outliers for the response variable, and compare the parameter estimates that are computed by least-squares and robust regression algorithms.

Exercise 11.15: Use PROC REG to run a regression analysis on the RegOutliers data. Include the CLB option in the MODEL statement to compute 95% confidence intervals for the parameters. Do the confidence intervals include the parameters for this example?

11.6.2   Simulating High-Leverage Points

Extreme values in the space of the explanatory variables are called high-leverage points. To simulate data that contain high-leverage points and outliers, do the following:

1. Generate a matrix of observations, X, from a multivariate contaminated normal distribution, as described in Section 8.5.1.

2. Generate error terms from a univariate contaminated normal distribution.

3. Form the regression model as $Y = X\beta + \epsilon$.

This algorithm is implemented in the following SAS/IML program. The explanatory variables are simulated from a multivariate normal distribution with 15% contamination. The error term is univariate normal with 25% contamination. In the simulation, the intercept term is assigned separately; the X matrix does not include a column of ones for the intercept. It is often more efficient not to waste space and computations on that scalar quantity.

```
/* simulate outliers and high-leverage points for regression data   */
%let N = 100;
proc iml;
call randseed(1);
mu =    {0 0 0};                          /* means                   */
Cov = {10 3 -2, 3 6 1, -2 1 2};           /* covariance for X        */
kX = 25;                                  /* contamination factor for  X */
pX = 0.15;                                /* prob of contamination for X */
kY = 10;                                  /* contamination factor for  Y */
pY = 0.25;                                /* prob of contamination for Y */

/* simulate contaminated normal (mixture) distribution */
call randgen(N1, "Binomial", 1-pX, &N); /* N1=num of uncontaminated */
X = j(&N, ncol(mu));
X[1:N1,] = RandNormal(N1, mu, Cov);          /* draw N1 from uncontam */
X[N1+1:&N,] = RandNormal(&N-N1, mu, kX*Cov); /* N-N1 from contam     */

/* simulate error term according to contaminated normal */
outlier = j(&N, 1);
call randgen(outlier, "Bernoulli", pY);    /* choose outliers        */
eps = j(&N, 1);
call randgen(eps, "Normal", 0, 1);         /* uncontaminated N(0,1) */
outlierIdx = loc(outlier);
if ncol(outlierIdx)>0 then                 /* if outliers...         */
   eps[outlierIdx] = kY * eps[outlierIdx]; /* set eps ~ N(0,kY)      */

/* generate Y according to regression model */
beta = {2, 1, -1};                         /* params, not including intercept */
Y = 1 + X*beta + eps;
```

The vector **Y** is the response vector for the regression model. In order to estimate the parameters in the model, you need to use a robust regression algorithm that detects high-leverage points. The following statements write the data to a SAS data set and call the ROBUSTREG procedure with the LTS algorithm to analyze the data. The LEVERAGE option in the MODEL statement requests the computation of leverage diagnostics. Both outliers and high-leverage points are written to the Out data set by the LEVERAGE= option in the OUTPUT statement.

```
/* write SAS data set */
varNames = ('x1':'x3') || {"Y" "Outlier"};
output = X || Y || outlier;
create Robust from output[c=varNames];  append from output;  close;
quit;

proc robustreg data=Robust method=LTS plots=RDPlot;
   model Y = x1-x3 / leverage;
   output out=Out outlier=RROutlier leverage=RRLeverage;
   ods select LTSEstimates DiagSummary RDPlot;
run;
```

Figure 11.12 Robust Parameter Estimates for Contaminated Data

The ROBUSTREG Procedure

LTS Parameter Estimates		
Parameter	DF	Estimate
Intercept	1	0.8793
x1	1	1.9874
x2	1	1.0194
x3	1	-1.0270
Scale (sLTS)	0	1.1153
Scale (Wscale)	0	0.9085

Diagnostics Summary		
Observation Type	Proportion	Cutoff
Outlier	0.1500	3.0000
Leverage	0.1800	3.0575

The parameter estimates are close to the parameter values. The "Diagnostics Summary" table reports how many outliers and high-leverage points were detected. For these simulated data, 15 (15% of the 100 observations) were classified as outliers and 18 points have high-leverage.

The classification of the observations is shown graphically in Figure 11.13. Points that are plotted outside of the horizontal lines at ± 3 are classified as outliers. Points that are to the right of the vertical line at 3.0575 are classified as high-leverage points. Outliers and high-leverage points are marked by their observation numbers. The high-leverage points correspond to observations near the end of the data set. (Recall that the first N_1 observations were drawn from an uncontaminated distribution.)

The indicator variables RROutlier and RRLeverage in the output data set identify which observations were classified as outliers and leverage points, respectively. You can compare these indicator variables with the simulated data.

Exercise 11.16: Use PROC FREQ to create a crosstabulation table that shows the relationship between the Outlier and RROutlier variables in the Out data set. The Outlier variable indicates how the observation was generated. Explain why there are observations with Outlier=1 that are not classified as outliers by the ROBUSTREG procedure.

Figure 11.13 Outliers and High-Leverage Points

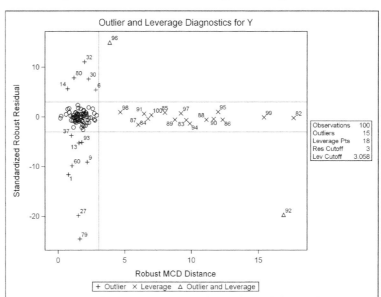

11.7 References

Greene, W. H. (2000), *Econometric Analysis*, 4th Edition, Upper Saddle River, NJ: Prentice-Hall.

Chapter 12
Simulating Data for Advanced Regression Models

Contents

12.1 Overview of Advanced Regression Models

Chapter 11, "Simulating Data for Basic Regression Models," describes how to simulate data from basic regression models for which the predicted response is a linear combination of explanatory variables. This chapter describes more complicated regression models for which the response variable is a more general function of the explanatory variables, plus random error. This chapter examines the following models:

- logistic and Poisson regression

- mixed models with random effects and correlated errors

- models that arise in survival analysis, such as the Cox proportional hazards model

- nonparametric regression models, such as loess regression

Mixed-effect models assume the existence of *random effects* that are distributed as a multivariate normal with zero mean and a specified variance-covariance matrix. The error term is also assumed to be multivariate normal. Section 10.4.4 describes how to construct a covariance matrix with a given structure.

For generalized linear models, a *link function* is used to model the mean of the response variable as a function of the *linear predictor*, which is a linear combination of the explanatory variables. For example, in a logistic regression the response variable is assumed to have a binary (Bernoulli) distribution, and a logistic function is used to link the linear predictor to the mean of the response. Poisson regression uses a logarithmic link function to model the mean of the response as a function of the linear predictor.

For an example of simulating data from a structural equations model, see Fan et al. (2002, Ch. 7).

12.2 Generalized Linear Regression Model

In a generalized linear model, the response variable follows a distribution whose parameters depend on a linear function of the covariates. A generalized linear model consists of three components:

- A linear predictor: $\eta = X\beta$

- A monotonic differentiable function, g, which is called a *link function*, that relates the expected value of the response to the linear predictor: $g(\mu_i) = X_i\beta$, where μ_i is the mean of Y_i and X_i is the ith row of X.

- A *variance function*, V, that relates the variance of Y_i to the mean μ_i. For some constant, ϕ, which is called the *dispersion constant*, the variance of Y_i is $\phi V(\mu_i)$.

Common choices for the link function include the identity function, the logit function, and the logarithm function. The response Y_i is assumed to follow a distribution in the *exponential family*, which includes the normal, gamma, binomial, Poisson, and negative binomial distributions. Notice that the binomial distribution includes the Bernoulli distribution as a special case.

12.2.1 Linear Regression Revisited

Section 11.5 describes how to simulate data from a linear regression model. In that section, the linear regression model is written as

$$Y_i = X_i\beta + \epsilon_i$$

for $i = 1, \ldots, N$, where X is an $N \times p$ design matrix whose ith row is X_i, β is a p-dimensional vector of regression parameters, and each ϵ_i is independently distributed as $N(0, \sigma)$.

This is equivalent to describing Y_i as being normally distributed with distribution $N(X_i\beta, \sigma)$. In other words, Y_i is a random value from the normal distribution with expected value $\eta_i = X_i\beta$ and constant variance $\phi = \sigma^2$. Consequently, a linear regression is a particular case of the generalized linear model. The link function is the identity function.

The following SAS/IML statements generate a vector of responses for linear regression:

```
proc iml;
...
eta = X*beta;
call streaminit(1);
y = rand(y, "Normal", eta, sigma);
```

The previous statements call the STREAMINIT subroutine and the RAND function, which until now have only been called in a DATA step and never in a SAS/IML program. However, you can call any DATA step function from PROC IML, and you can pass in a vector of parameters. Notice that `eta` is an $N \times 1$ vector and that the goal is to generate N response values, each from a different distribution. Prior to SAS/IML 9.3, the RANDGEN routine did not support vectors of parameter values. That feature was added in SAS/IML 12.1.

Exercise 12.1: Fill in the details in the PROC IML program and simulate 1,000 observations from the univariate generalized linear model, where Y_i is normally distributed with mean $1 - 2X_i$ and variance 0.5. Run PROC REG to check the simulation.

12.2.2 Logistic Regression Model

An example of a nontrivial link function is provided by a logistic regression model. For a logistic model, the logit transformation is used to transform the linear predictor to the range $[0, 1]$. If you define the link function by $\eta_i = g(\mu_i) = \log(\mu_i/(1 - \mu_i))$, a little algebra inverts the expression to give $\mu_i = \exp(\eta_i)/(1 + \exp(\eta_i))$, which is called the logistic transformation of η_i.

The response variable in a logistic regression is assumed to be a Bernoulli variable: $Y_i \sim$ Bernoulli(μ_i). The choice of a Bernoulli distribution determines the expected value (and variance) of Y_i: $E(Y_i) = \mu_i$. The following DATA step simulates data for a logistic regression model. Notice that the definition of η does *not* include a random term. Instead, the randomness is achieved by calling the RAND function to generate y as a Bernoulli random variate with parameter $\mu(\eta)$.

```
%let N = 150;
data LogisticData;
array xx1{&N} _temporary_;
array xx2{&N} _temporary_ ;
call streaminit(1);

/* read or simulate fixed effects */
do i = 1 to &N;
   xx1{i} = rand("Uniform");   xx2{i} = rand("Normal", 0, 2);
end;

/* simulate logistic model */
do i = 1 to &N;
   x1 = xx1{i};   x2 = xx2{i};
   /* linear model with parameters  {2, -4, 1} */
   eta = 2 - 4*x1 + 1*x2;              /* eta = X*beta. NO epsilon!  */
   mu = exp(eta) / (1+exp(eta));       /* transform by inverse logit */
   y = rand("Bernoulli", mu);          /* binary response            */
   output;
end;
run;
```

You can verify that the simulated data are reasonable by running the LOGISTIC procedure, as follows. The output is shown in Figure 12.1.

```
ods graphics on;
proc logistic data=LogisticData plots(only)=Effect;
   model y(Event='1') = x1 x2 / clparm=wald;
   ods select ParameterEstimates CLParmWald EffectPlot;
run;
```

Figure 12.1 Parameter Estimates for a Logistic Model

The LOGISTIC Procedure

Analysis of Maximum Likelihood Estimates					
Parameter	DF	Estimate	Standard Error	Wald Chi-Square	Pr > ChiSq
Intercept	1	2.0111	0.5140	15.3099	<.0001
x1	1	-3.4947	0.8644	16.3431	<.0001
x2	1	0.8785	0.1524	33.2066	<.0001

Parameter Estimates and Wald Confidence Intervals			
Parameter	Estimate	95% Confidence Limits	
Intercept	2.0111	1.0037	3.0185
x1	-3.4947	-5.1890	-1.8004
x2	0.8785	0.5797	1.1773

There are only 150 observations in the simulated data, so the parameter estimates are not very close to the values of the parameters. However, the 95% Wald confidence intervals for the parameters do include the true values of 2, -4, and 1.

In Figure 12.2, the fitted model is overlaid on the graph of the response versus the x1 variable (at the mean value of x2). About half of the generated responses are zeros, and about half of the generated responses are ones. This occurred because the parameter values for the model were chosen so that the median value of mu is about 0.5. Different parameter values might lead to a less symmetric distribution of responses.

Exercise 12.2: Run the logistic simulation with $N = 1,000$ observations. Compare the standard errors to those shown in Figure 12.1.

Exercise 12.3: The Wald confidence intervals for parameters are based on asymptotic normality of the parameter estimates, which is not accurate for small sample sizes. Compare the Wald intervals with the empirical 95% quantiles of the sampling distributions when the sample size is 100 observations. (Remember to use the %ODSOFF macro to suppress output, and do not use the PLOTS= option.)

Figure 12.2 Logistic Data and Model

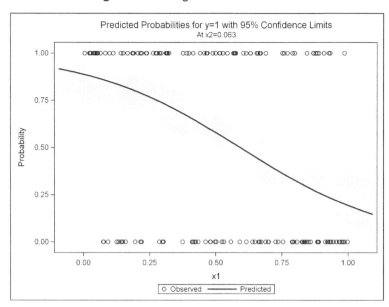

12.2.3 Poisson Regression Model

A Poisson regression model is another example of a generalized linear model with a nontrivial link function. For a Poisson regression model, the logarithmic transformation is used to relate the linear predictor to the mean of the Poisson distribution. If you define the link function by $\eta_i = g(\mu_i) = \log(\mu_i)$, then the inverse transformation is $\mu_i = \exp(\eta_i)$.

The response variable is assumed to follow a Poisson distribution: $Y_i \sim \text{Poisson}(\mu_i)$. Therefore, the expected value (and variance) of Y_i is μ_i. You can use the DATA step to simulate data for a Poisson regression model. The program is identical to the program in Section 12.2.2 except for the definition of the variable mu and the fact that y is drawn from a Poisson distribution. Again, notice that there is not a random term associated with η. Randomness is achieved by calling the RAND function to generate y as a Poisson random variate with parameter $\mu(\eta)$.

```
eta = 2 - 4*x1 + 1*x2;          /* eta = X*beta           */
mu = exp(eta);                  /* transform linear predictor */
y = rand("Poisson", mu);        /* Poisson response       */
```

You can use the GENMOD procedure to fit a Poisson model to the simulated data. Figure 12.3 shows that the parameter estimates are close to the parameter values and that the Wald confidence intervals include the parameter values.

```
proc genmod data=PoissonData;
   model y = x1 x2 / dist=Poisson;
   ods select ParameterEstimates;
run;
```

Figure 12.3 Parameter Estimates for a Poisson Model

The GENMOD Procedure

Analysis Of Maximum Likelihood Parameter Estimates							
Parameter	DF	Estimate	Standard Error	Wald 95% Confidence Limits		Wald Chi-Square	Pr > ChiSq
Intercept	1	2.0959	0.0952	1.9093	2.2826	484.36	<.0001
x1	1	-4.2554	0.1629	-4.5747	-3.9362	682.52	<.0001
x2	1	0.9715	0.0289	0.9149	1.0282	1131.14	<.0001
Scale	0	1.0000	0.0000	1.0000	1.0000		

Note: The scale parameter was held fixed.

12.3   Linear Mixed Models

Previous sections have simulated data from a linear regression model with fixed effects. That is, the response variable is modeled by $Y_i = X_i \beta + \epsilon_i$, or, in matrix-vector notation, $\mathbf{Y} = \mathbf{X}\beta + \epsilon$.

A *linear mixed model* adds *random effects* to the model. In matrix notation, a linear mixed model can be written as

$$\mathbf{Y} = \mathbf{X}\beta + \mathbf{Z}\gamma + \epsilon$$

where

$\mathbf{Y}$ is the vector of observed responses.

$\mathbf{X}$ is the known design matrix of fixed effects.

β is a vector of fixed (but unknown) parameters.

$\mathbf{Z}$ is the known design matrix for the random effects.

γ is the vector of random effects.

ϵ is a vector of random errors.

The textbook assumptions for linear mixed models are as follows:

- The random effects γ are assumed to follow a multivariate normal distribution with mean $\mathbf{0}$ and covariance matrix $\mathbf{G}$.

- The random errors ϵ are assumed to follow a multivariate normal distribution with mean $\mathbf{0}$ and covariance matrix $\mathbf{R}$. The random errors are independent of the random effects.

Section 10.4.4 describes several matrices that are frequently used to model the random errors in mixed models. Notice that if $\mathbf{Z} = \mathbf{0}$ and $\mathbf{R} = \sigma^2 \mathbf{I}$, then the mixed model reduces to the general linear model.

Notice also that you can use a computational trick to generate a random variable that follows a multivariate normal distribution with a diagonal covariance matrix. If $\epsilon \sim \text{MVN}(\mathbf{0}, \sigma^2 \mathbf{I})$, then $\epsilon_i \sim N(0, \sigma)$. This trick generalizes. Accordingly, you can use the DATA step to generate random effects for any diagonal covariance matrix. For more complicated covariance structures, use the SAS/IML language.

Although this section uses a boldface font to emphasize that the equations apply to matrices and vectors, most of this book does not use boldface, but relies on context to indicate that a symbol represents a nonscalar quantity.

12.3.1   A Repeated Measures Model with a Random Effect

As a first example, consider a simple regression model that has a single random effect and no fixed effects. Suppose that the random effect is a classification variable with L levels, and for each level of the random effect there are k observations. The random effect model for repeated measurements is

$$Y_{ij} = \mu + A_i + \epsilon_{ij}$$

where μ is an overall mean, $A_i \sim N(0, \sigma_A)$ and $\epsilon_{ij} \sim N(0, \sigma)$ for $i = 1, \ldots, L$ and $j = 1, \ldots, k$.

The following DATA step simulates data for this regression model when $\mu = 5$. The number of classification levels and the number of repeated measurements are provided as macro parameters.

```
/* simple random effect model with repeated measurements */
%let var_A  = 4;     /* variance of random effect (intercept)        */
%let sigma2 = 2;     /* variance of residual, e ~ N(0, sqrt(sigma2)) */
%let L = 3;          /* num levels in random effect A                */
%let k = 5;          /* num repeated measurements in each level of A */
data RandomEffects(drop=mu rndA);
call streaminit(12345);
mu = 5;
do a = 1 to &L;
   rndA = rand("Normal", 0, sqrt(&var_A));
   do rep = 1 to &k;
      y = mu + rndA + rand("Normal", 0, sqrt(&sigma2));
      output;
   end;
end;
run;
```

You can use the MIXED procedure to validate the properties of the simulated data. The results of the PROC MIXED analysis are shown in Figure 12.4.

```
proc mixed data=RandomEffects CL;
   class a;
   model y = ;                      /* no fixed effects             */
   random int / subject=a;          /* a is random intercept effect */
   ods select CovParms;
run;
```

Figure 12.4 Parameter Estimates for a Repeated Measures Model

The Mixed Procedure

Covariance Parameter Estimates					
Cov Parm	Subject	Estimate	Alpha	Lower	Upper
Intercept	a	3.2325	0.05	0.8213	212.90
Residual		1.2461	0.05	0.6408	3.3955

The CovParms table shows estimates of the two variance terms in the model. The "Intercept" row estimates the variance of the random intercept term, which is 4 in this example. The "Residual" row estimates the variance of the random error term, which is 2. Notice that the confidence intervals are quite wide because the sample is so small. The following exercise examines the sampling distribution of these estimates.

Exercise 12.4: Create 1,000 sets of simulated data by using an additional DO loop in the DATA step. Use a BY statement in the MIXED procedure to estimate the sampling distribution of the variance of the random effect and of the error term. Are the Monte Carlo estimates close to 4 and 2, respectively?

12.3.2   Simulating Correlated Random Errors

This section presents a repeated measures example that is similar to the example called "Growth Curve with Compound Symmetry" in the documentation for the MIXED procedure in the *SAS/STAT User's Guide*. Suppose that five individuals start a diet plan and that the change in their weight is measured every week for three weeks. To fit a linear trend in time to each individual, the X matrix is as follows:

$$X = \begin{bmatrix} 1 & 1 \\ 1 & 2 \\ 1 & 3 \\ \vdots & \vdots \\ 1 & 1 \\ 1 & 2 \\ 1 & 3 \end{bmatrix}$$

You might want to model correlation among repeated measurements from the same individual by using the R matrix and setting the Z matrix to zero. For this example, you can construct a 15×15 block-diagonal matrix R where each 3×3 block has compound symmetry. This structure has two covariance parameters: σ_{CS}^2 is the common covariance and σ_R^2 is the residual covariance. The R matrix is as follows:

$$
R = \begin{bmatrix}
\sigma_{CS}^2 + \sigma_R^2 & \sigma_R^2 & \sigma_R^2 & & & & \\
\sigma_R^2 & \sigma_{CS}^2 + \sigma_R^2 & \sigma_R^2 & & & & \\
\sigma_R^2 & \sigma_R^2 & \sigma_{CS}^2 + \sigma_R^2 & & & & \\
& & & \ddots & & & \\
& & & & \sigma_{CS}^2 + \sigma_R^2 & \sigma_R^2 & \sigma_R^2 \\
& & & & \sigma_R^2 & \sigma_{CS}^2 + \sigma_R^2 & \sigma_R^2 \\
& & & & \sigma_R^2 & \sigma_R^2 & \sigma_{CS}^2 + \sigma_R^2
\end{bmatrix}
$$

12.3.2.1   Constructing a Block-Diagonal Matrix

There are several ways to construct a block-diagonal matrix in SAS/IML software. For this example, you could explicitly type the matrix, but it is better to learn how to construct the R matrix for the general case of s individuals, each of which is associated with k measurements.

The first technique is to use the BLOCK function in the SAS/IML language. The expression `block(B1,B2,...)` returns a block-diagonal matrix with the specified matrices `B1, B2,...` along the block diagonal. The matrices do not have to be the same size, but they are for this application.

If you know that $s = 5$, you can construct the 3×3 compound-symmetric matrix, B, and assign `R = block(B,B,B,B,B)`. However, the general case requires a loop with s iterations, as shown in the following statements:

```
proc iml;
k=3;                              /* 3 measurements on growth curve  */
s=5;                              /* 5 individuals                   */
sigma2_R = 1.4;                   /* Parameter 1: residual covariance */
sigma2_CS = 2;                    /* Parameter 2: common covariance   */
B = sigma2_R*j(k,k,1) + sigma2_CS*I(k);      /* cs matrix            */
R = B;                            /* first block                     */
do i = 2 to s;                    /* create block-diagonal matrix    */
   R = block(R, b);               /*    with s blocks                */
end;
```

There is a second way to construct a block-diagonal matrix that does not require a loop, but uses the *Kronecker product* matrix operator. If $A = \{A_{ij}\}$ is an $s \times s$ matrix and B is a $k \times k$ matrix, then the Kronecker product $A \otimes B$ is the following $sk \times sk$ block matrix:

$$
A \otimes B = \begin{bmatrix}
A_{11}B & \cdots & A_{1n}B \\
\vdots & \ddots & \vdots \\
A_{n1}B & \cdots & A_{nn}B
\end{bmatrix}
$$

In particular, if I_s is an $s \times s$ identity matrix, then $I_s \otimes B$ is block diagonal. The direct product operator in the SAS/IML language is represented by using the @ symbol. Consequently, an equivalent way to specify a block-diagonal matrix is by using the following statement:

```
R = I(s) @ B;                            /* block-diagonal matrix */
```

12.3.2.2   Simulating the Mixed Model

After the R matrix is defined, you can use the SAS/IML language to simulate responses that satisfy the model

$$Y = X\beta + \epsilon$$

where $\epsilon \sim N(0, R)$. (This is a matrix-vector equation.) Assume a mean weight loss of 2 pounds (0.9 kg) per week. Also assume that $\sigma^2_{CS} = 2$ and $\sigma^2_R = 1.4$. The following program (which continues from the previous section) simulates a single set of 15 observations:

```
beta = {0, -2};                    /* parameters for fixed effects */
Week = T(repeat(1:k, 1,s));        /* column: 1,2,3,1,2,3,...      */
X = j(nrow(Week),1,1) || Week;     /* add intercept                */
Indiv = colvec(repeat(T(1:s),1,k)); /* 1,1,1,2,2,2,3,3,3,...       */

call randseed(1234);
create Mix var {"WtLoss" "Week" "Indiv"};      /* name the variables */

/* random effects */
zero = j(1, k*s, 0);               /* the zero vector              */
eps = RandNormal(1, zero, R);      /* eps ~ MVN(0,R)               */
WtLoss = X*beta + eps`;            /* fixed effects, correlated errors */

append;
close;
quit;
```

It is always a good idea to graph the data to see whether the simulation appears to be correct. Figure 12.5 should give you confidence that the fixed effects are specified correctly because each individual is losing about 2 pounds per week.

```
proc sgplot data=Mix;
   label WtLoss = "Weight Loss";
   series x=Week y=WtLoss / group=Indiv markers;
   refline 0 / axis=y;
   xaxis integer;
run;
```

Figure 12.5 Model of Weight Loss

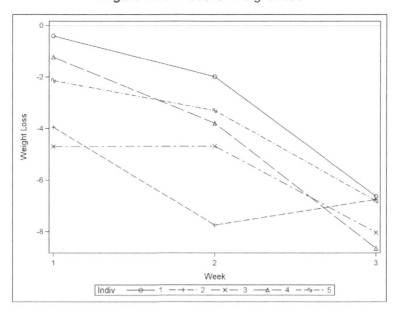

Figure 12.5 does not reveal whether the error terms exhibit a compound symmetric correlation structure. To see that, you have to use PROC MIXED to fit the model to the simulated data and compute the covariance of the residuals. Because the parameter estimates might not be close to the parameters, it is useful to use the CL option in the PROC MIXED statement to compute (Wald) confidence intervals for the covariance parameters. Figure 12.6 shows that for these simulated data, the confidence intervals include the parameter values $\sigma_{CS}^2 = 2$ and $\sigma_R^2 = 1.4$. It is therefore plausible that the simulation program is implemented correctly.

```
proc mixed data=Mix CL;
   class Indiv;
   model WtLoss = Week;
   repeated / type=cs subject=Indiv R;
   ods select Dimensions R CovParms;
run;
```

Figure 12.6 Parameter Estimates for a Mixed Effects Model

The Mixed Procedure

Dimensions	
Covariance Parameters	2
Columns in X	2
Columns in Z	0
Subjects	5
Max Obs Per Subject	3

Figure 12.6 *continued*

Estimated R Matrix for Indiv 1			
Row	Col1	Col2	Col3
1	2.8791	1.0324	1.0324
2	1.0324	2.8791	1.0324
3	1.0324	1.0324	2.8791

Covariance Parameter Estimates					
Cov Parm	Subject	Estimate	Alpha	Lower	Upper
CS	Indiv	1.0324	0.05	-1.3213	3.3861
Residual		1.8467	0.05	0.8737	6.1547

Exercise 12.5: This exercise examines the distribution of the covariance estimates for the preceding mixed model. Complete the following steps:

1. Use the following SAS/IML statements to generate 1,000 samples for the mixed model:

```
call randseed(1234);
create Mix var {"SampleID" "WtLoss" "Week" "Indiv"};

/* random effects */
do i = 1 to 1000;
   eps = RandNormal(1, zero, R);
   WtLoss = X*beta + eps`;
   SampleID = repeat(i, k*s);
   append;
end;
close;
```

2. Run PROC MIXED and use a BY SampleID statement to generate 1,000 parameter estimates by writing the CovParms table to a SAS data set. (Hint: This is a good time to use `options nonotes` as well as the %ODSOFF macro.)

3. Use PROC MEANS and PROC UNIVARIATE to describe the distribution of the covariance parameter estimates.

12.3.3   Simulating Random Effects Components

A more complicated mixed model might include fixed and random effects. For example, the following data and analysis are included in the "Getting Started" section of the MIXED procedure. The data are heights in inches, which are measured for multiple individuals within four families. The researcher assumes that the heights can be modeled by using gender as a fixed effect. The researcher includes "family" as a random effect because she assumes that heights are correlated between members of a family. The researcher also includes an interaction term between gender and family in her model as a random effect. The data and analysis are shown in the following example; the parameter estimates are shown in Figure 12.7.

```
/* Getting Started example in PROC MIXED documentation */
data heights;
  input Family Gender$ Height @@;
  datalines;
1 F 67   1 F 66   1 F 64   1 M 71   1 M 72
2 F 63   2 F 63   2 F 67   2 M 69   2 M 68   2 M 70
3 F 63   3 M 64
4 F 67   4 F 66   4 M 67   4 M 67   4 M 69
;
run;

/* Model data. Save parameter estimates */
proc mixed data=heights;
  class Family Gender;
  model Height = Gender / solution outpm=outpm;
  random Family Family*Gender;
  ods select CovParms SolutionF;
  ods output CovParms=CovParms SolutionF=SolutionF;
run;
```

Figure 12.7 Parameter Estimates for a Mixed Effects Model

The Mixed Procedure

Covariance Parameter Estimates	
Cov Parm	Estimate
Family	2.4010
Family*Gender	1.7657
Residual	2.1668

Solution for Fixed Effects						
Effect	Gender	Estimate	Standard Error	DF	t Value	Pr > |t|
Intercept		68.2114	1.1477	3	59.43	<.0001
Gender	F	-3.3621	1.1923	3	-2.82	0.0667
Gender	M	0	.	.	.	.

The PROC MIXED statements use dummy variables that are associated with the random effects to construct the Z matrix in the mixed model. The matrix of fixed effects, X, contains a column of 1s for the intercept term as well as dummy variables for the Gender variable.

Suppose that you want to simulate new data for this fitted model. That is, you want to simulate the height of 18 individuals from four families, where the heights are generated according to the mixed model whose estimates are shown in Figure 12.7. Because you do not know the values for the parameters, you might choose to use the values of the parameter estimates. To simulate additional samples, do the following:

1. Create the matrix of fixed and random effects that correspond to the design of the data. You can use the GLMMOD or GLIMMIX procedures to do this.

2. Construct the G matrix, which is a diagonal matrix that contains the variance components for the Family and Family*Gender effects.

3. Construct the R matrix, which for this example is simply $R = \sigma^2 I$, where σ^2 is the variance of the error term.

4. For each simulated individual, compute the height of that individual as the sum of the following terms:

   a) The linear predictor of the fixed effects, $\eta = X\beta$. The parameter estimates in the SolutionF data set are used instead of the unknown parameters.

   b) The random effects, which are $Z\gamma$, where $\gamma \sim \text{MVN}(0, G)$.

   c) The error term, ϵ, where $\epsilon \sim \text{MVN}(0, R)$. Equivalently, since $R = \sigma^2 I$, $\epsilon_i \sim N(0, \sigma)$.

This algorithm requires matrices, vectors, and multivariate distributions. Consequently, the SAS/IML language is the easiest way to implement it. The four steps are implemented in the next sections.

12.3.3.1   Create Design Matrices for Fixed and Random Effects

There are two ways to create the design matrices for the fixed and random effects. The first way is to call the GLMMOD procedure twice, once for the fixed effects and once for the random effects. The second way is to call the GLIMMIX procedure, and use the OUTDESIGN= option to output a single design matrix.

The GLIMMIX procedure enables you to generate the X and Z matrices in a single call. Unlike the LOGISTIC procedure (see Section 11.5.2.2), the GLIMMIX procedure does not support the OUTDESIGNONLY option as of SAS 9.3. Therefore, the GLIMMIX procedure fits a mixed model in addition to generating the design matrices. The following statements create the design matrices in the All data set. Information about the design matrices is shown in Figure 12.8.

```
proc glimmix data=heights outdesign(names novar)=All;
   class Family Gender;
   model Height = Gender;
   random Family Family*Gender;
   ods select ColumnNames;
run;
```

The preceding call to the GLIMMIX procedure does the following:

- The OUTDESIGN= option creates the All data set, which contains 18 observations and the dummy variables _X1–_X3 and _Z1–_Z12.

- The NOVAR option to the OUTDESIGN= option excludes the input variables from the All data set. Without this option, the data set would include a copy of the Family, Gender, and Height variables.

- The NAMES option to the OUTDESIGN= option creates the "ColumnNames" tables, shown in Figure 12.8, which associates each effect in the model with a column in the All data set.

Figure 12.8 Column Names for Design Matrices

The GLIMMIX Procedure

X Matrix Column Definitions		
Column	Effect	Gender
1	Intercept	
2	Gender	F
3	Gender	M

Z Matrix Column Definitions			
Column	Effect	Gender	Family
1	Family		1
2	Family		2
3	Family		3
4	Family		4
5	Family*Gender	F	1
6	Family*Gender	M	1
7	Family*Gender	F	2
8	Family*Gender	M	2
9	Family*Gender	F	3
10	Family*Gender	M	3
11	Family*Gender	F	4
12	Family*Gender	M	4

You can also use the GLIMMIX procedure to generate nonsingular parameterizations of the classification effects, as described in Section 11.5.2.2.

Exercise 12.6: Call the GLMMOD procedure twice to generate the two design matrices.

12.3.3.2 Read Design Matrices and Parameter Estimates

The GLIMMIX procedure outputs both design matrices in a single data set. The columns for fixed effects begin with the prefix "_X", whereas the columns for random effects begin with the prefix "_Z". The following SAS/IML statements read the X and Z matrices into PROC IML:

```
proc iml;
FixedVar = "_X1":"_X3";
RandomVar = "_Z1":"_Z12";
use All;
read all var FixedVar into X;
read all var RandomVar into Z;
close All;
```

You could manually specify the parameters for the fixed and random effects, but suppose that you intend to use the parameter estimates that were saved in Section 12.3.3. The parameter estimates for the fixed effects are contained in the SolutionF data set. You can read the estimates and construct the linear predictor as follows:

```
/* read parameter estimates for fxed effects into beta */
use SolutionF; read all var {Estimate} into beta; close;
eta = X*beta;                           /* linear predictor */
```

12.3.3.3   Construct the *G* and *R* Matrices

The parameter estimates for the variances of the random effects and for the error term are contained in the CovParms data set. The following SAS/IML statements read the estimates:

```
/* read estimates for covariance parameters */
use CovParms; read all var {Estimate} into var; close;
varF   = var[1];                       /* Var(Family)        */
varFG  = var[2];                       /* Var(Family*Gender) */
sigma2 = var[3];                       /* sigma2 = Var(Error) */
```

The *G* matrix for this example is a diagonal matrix. As shown in Figure 12.8, the first four variance components are for the Family variable, whereas the last eight are for the Family-Gender interaction. You can use the DIAG function to construct a diagonal matrix from these variance components, as follows:

```
/* Define covariance matrix G for random effects */
G = diag( repeat(varF,4) // repeat(varFG,8) );
```

The *R* matrix is a multiple of the identity matrix. Accordingly, instead of explicitly forming the *R* matrix and generating $\epsilon \sim \text{MVN}(0, R)$, the next section generates each component of ϵ as independent univariate draws: $\epsilon_i \sim N(0, \sigma)$.

12.3.3.4   Simulate Reponses

With the preliminary computations completed, you can now simulate the response variable according to the mixed model. Suppose that you want to simulate five new sets of data that have properties that are similar to the original data. The following SAS/IML statements show one way to simulate the responses:

```
call randseed(1);
zero = repeat(0, nrow(G));             /* the zero vector            */
NumSamples = 5;
```

```
/* first attempt (less efficient) */
eps = J(nrow(X),1);                        /* allocate error term      */
Y = j(nrow(X), NumSamples);                /* allocate cols for responses */
do j = 1 to NumSamples;                    /* simulate mixed model     */
   gamma = RandNormal(1, zero, G);              /* gamma ~ MVN(0,G) */
   call randgen(eps, "Normal", 0, sqrt(sigma2)); /* eps ~ N(0,sigma))*/
   Y[,j] = eta + Z*gamma` + eps;
end;
```

The SAS/IML code is a good first attempt to simulate the data. It works, and it uses matrix-vector computations for efficiency. However, whenever you see a SAS/IML loop, you should ask yourself whether there is a way to vectorize the program further. In this case, the answer is "yes." You can generate all of the random variates in a single call and use matrix-matrix operations to compute all of the responses:

```
/* second attempt (more efficient) */
gamma = RandNormal(NumSamples, zero, G);   /* each column is MVN(0,G) */
eps = J(nrow(X), NumSamples);                      /* allocate error term */
call randgen(eps, "Normal", 0, sqrt(sigma2)); /* eps[i,j]~N(0,sigma) */
Y = eta + Z*gamma` + eps;
```

The simulation is complete. However, it is always a good idea to check that the results make sense. You can visualize these simulated response variables by plotting each simulated sample, as follows. The result is shown in Figure 12.9.

```
yNames = "y1":("y"+strip(char(NumSamples)));              /* "y1":"y5" */
create Sim from Y[c=yNames];  append from Y;  close;
quit;

/* add subject identifier */
data Sim;
merge OutPM Sim;   /* merge simulated responses and predicted means */
N=_N_;
run;

proc sgplot data=Sim nocycleattrs;
   /* trick: Use scatter plot to show markers by gender */
   series x=N y=y1; scatter x=N y=y1 / group=Gender;
   series x=N y=y2; scatter x=N y=y2 / group=Gender;
   series x=N y=y3; scatter x=N y=y3 / group=Gender;
   series x=N y=y4; scatter x=N y=y4 / group=Gender;
   series x=N y=y5; scatter x=N y=y5 / group=Gender;

   /* show means for females and males */
   refline 64.85 68.21 / axis=y lineattrs=(pattern=dash);
   refline 5.5 11.5 13.5 / axis=x;              /* separate families */
   xaxis values=(1 to 18) label="Subjects";
   yaxis label="Height (in)";
run;
```

Figure 12.9 Variation of Response Around Mean

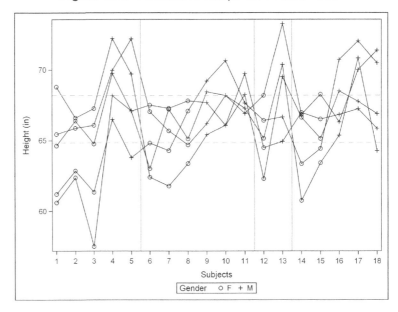

In Figure 12.9, a series plot is used to show each of the five simulated samples. Each series plot represents a single realization of simulated heights for the 18 individuals whose heights vary according to a mixed model that incorporates family and gender. The horizontal lines represent the predicted means for males (top line) and females (bottom line). The vertical lines separate families. The simulation exhibits random variation about the predicted heights, with the size of the variance depending on the family and on gender within the families.

Exercise 12.7: Simulate 100 new sets of data. Create box plots that show how the simulated heights vary for each Family-Gender combination.

12.4   Survival Analysis Models

Section 7.6 presents examples of simulating survival times. This section describes how to simulate survival data that include censoring and explanatory variables.

12.4.1   Proportional Hazards Model

A proportional hazards model assumes the existence of a baseline hazard function, $\lambda_0(t)$, which describes the survival distribution at the zero values of the explanatory variables. The hazard function for an individual is modeled as

$$\lambda_i(t; x_i) = \lambda_0(t) \exp(x_i' \beta)$$

where x_i is a vector of covariates for the ith individual, and β is a vector of regression coefficients. The exponential model and the Weibull model are proportional hazards models. For the Cox model, the baseline hazard function is unspecified.

As with other regression models, you can use simulation to study the distribution of the parameter estimates. Most simulation studies of the Cox regression model assume that the survival times follow an exponential or Weibull distribution (Bender, Augustin, and Blettner 2005, p. 1714). An exponential distribution of survival times occurs when the baseline hazard function is constant. This simple assumption is used in the following simulation, but Bender, Augustin, and Blettner (2005, p. 1715) discuss "how survival times can be generated to simulate Cox models with known regression coefficients and with any non-zero baseline hazard rate."

Recall from Section 2.5.3 that the following macro enables you to simulate data from an exponential distribution with scale parameter `sigma`:

```
%macro RandExp(sigma);
   ((&sigma) * rand("Exponential"))
%mend;
```

For survival models, the more useful exponential parameter is the "rate parameter," which is `1/sigma`.

In the following simulation (which is based on an example in Kleinman and Horton (2010)), the macro is used to make it clear that the survival time and the censoring time are exponentially distributed:

```
%let N = 100;
data PHData(keep=x1 x2 t censored);
array xx1{&N} _temporary_;
array xx2{&N} _temporary_ ;
call streaminit(1);

/* read or simulate fixed effects */
do i = 1 to &N;
   xx1{i} = rand("Normal");   xx2{i} = rand("Normal");
end;

/* simulate regression model */
baseHazardRate = 0.002;   censorRate = 0.001;
do i = 1 to &N;
   x1 = xx1{i};   x2 = xx2{i};
   eta = -2*x1 + 1*x2;                  /* form the linear predictor    */

   /* construct time of event and time of censoring */
   tEvent = %RandExp( 1/(baseHazardRate * exp(eta)) );
   c   = %RandExp( 1/censorRate );   /* rate parameter = censorRate   */
   t   = min(tEvent, c);              /* time of event or censor time  */
   censored = (c < tEvent);           /* indicator variable: censored? */
   output;
end;
run;
```

In the simulation, two continuous explanatory variables are simulated. The regression parameters are -2 and 1. The hazard function for an individual with $x_1 = x_2 = 0$ is the baseline hazard function, which is constant. The linear predictor is computed for each individual, and is used to simulate an event time.

The model also incorporates censoring. In this model, censoring occurs at a constant rate. If the time of censoring occurs before the event occurs, then the event is censored. An indicator variable, censored, specifies whether the time to the event is censored for an individual. In SAS, proportional hazards models are fit by using the PHREG procedure. You can use the PHREG procedure to estimate the regression coefficients and to estimate the survival function at specific covariate values for these simulated data, as follows:

```
ods graphics on;
proc phreg data=PHData plots(overlay CL)=(Survival);
   model t * censored(1) = x1-x2;
   ods select CensoredSummary ParameterEstimates
           ReferenceSet SurvivalPlot;
run;
```

The results are shown in Figure 12.10 and Figure 12.11. The simulation produces 40 censored observations. The parameter estimates are close to the parameter values. Figure 12.11 is an estimate of the baseline survival function.

Figure 12.10 Summary and Parameter Estimates for Proportional Hazards Regression

The PHREG Procedure

Summary of the Number of Event and Censored Values			
Total	Event	Censored	Percent Censored
100	60	40	40.00

Analysis of Maximum Likelihood Estimates						
Parameter	DF	Parameter Estimate	Standard Error	Chi-Square	Pr > ChiSq	Hazard Ratio
x1	1	-1.89107	0.22499	70.6436	<.0001	0.151
x2	1	0.98754	0.15519	40.4935	<.0001	2.685

Reference Set of Covariates for Plotting	
x1	x2
-0.000791	0.056205

Figure 12.11 Plot of the Estimated Survival Function

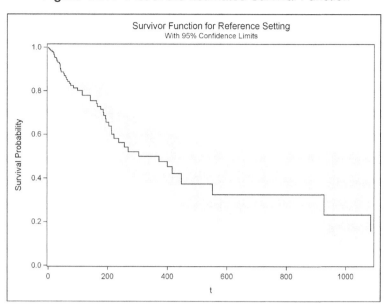

12.4.2   Simulating Data from Multiple Survivor Functions

When analyzing two or more samples of survival data, an important task is determining whether the underlying populations have identical survival functions. The survivor function (or *survival distribution function* (SDF)) at a time t is the probability that a randomly selected observation from the population will have a lifetime, T, that exceeds t. In symbols, the survivor function is $S(t) = P(T > t)$.

The following DATA step simulates survival data from two populations that have different survivor functions. For simplicity, the model assumes that a subject is censored with probability 0.2. The program uses an exponential distribution for the survival time.

```
%let N = 100;
data survsamp(keep=Treatment t Censored);
call streaminit(1);
array rate{2} (0.05 0.08);
do Treatment = 1 to dim(rate);
   do i = 1 to &N;
      censored = rand("Bernoulli", 0.2);
      t = %RandExp( 1/rate{Treatment} );
      output;
   end;
end;
run;
```

The LIFETEST procedure can test whether two samples are likely to have come from the same survivor function. The following statements test the simulated data for the two treatment groups:

```
ods graphics on;
proc lifetest data=survsamp plots=(survival);
   strata Treatment;
   time t * censored(1);
   ods select Quartiles HomTests SurvivalPlot;
run;
```

Figure 12.12 Quartiles and Tests of Equality

The LIFETEST Procedure

Stratum 1: Treatment = 1

		Quartile Estimates		
			95% Confidence Interval	
Percent	Point Estimate	Transform	[Lower	Upper)
75	33.230	LOGLOG	24.542	43.381
50	15.635	LOGLOG	12.962	20.699
25	6.902	LOGLOG	3.393	11.242

The LIFETEST Procedure

Stratum 2: Treatment = 2

		Quartile Estimates		
			95% Confidence Interval	
Percent	Point Estimate	Transform	[Lower	Upper)
75	27.346	LOGLOG	15.361	31.192
50	10.362	LOGLOG	7.229	12.113
25	4.767	LOGLOG	2.950	6.210

The LIFETEST Procedure

Test of Equality over Strata			
Test	Chi-Square	DF	Pr > Chi-Square
Log-Rank	7.4581	1	0.0063
Wilcoxon	6.0136	1	0.0142
-2Log(LR)	8.1252	1	0.0044

Figure 12.12 indicates that the median survival time for the first treatment group is 15.6. (The units—days, months, years—depend on the units for the rate function.) The median survival time for the second treatment group is 10.4. Each of the tests for equality indicate that it is unlikely that these

two samples come from the same survivor function. The estimated survivor functions are shown in Figure 12.13.

Figure 12.13 Estimated Survival Functions

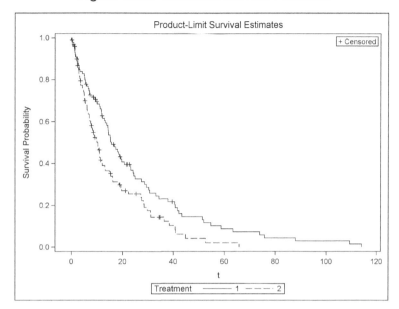

12.5 Nonparametric Models

You can use simulation to assess how well a nonparametric method fits data that are generated from a known function that has been corrupted by random noise. The technique is as follows:

1. Generate the response variable as a known function of the explanatory variables plus a random error term: $y_i = f(x_i) + \epsilon_i$, where $\epsilon_i \sim N(0, \sigma)$.

2. Use a nonparametric regression procedure (such as the LOESS, TPSPLINE, or GAM procedures) to fit a model, $\hat{y}_i = g(x_i)$, to the data.

3. Measure how far the model is from the true relationship by computing some quantity such as the root mean square error: $\sqrt{\Sigma_i (f(x_i) - g(x_i))^2 / N}$

As an example, the following DATA step generates a response variable given by $y_i = \sin(x_i/5) + 0.2 \cos(x_i) + \epsilon_i$ for a range of values of x_i, where $\epsilon_i \sim N(0, 0.2)$. A scatter plot of the data and a loess fit is shown in Figure 12.14.

```
data NonParam;
call streaminit(1);
do x = 1 to 30 by 0.1;
   f = sin(x/5) + 0.2*cos(x);
   y = f + rand("Normal", 0, 0.2);
   output;
end;
```

```
ods graphics on;
proc loess data=NonParam;
   model y = x;
   score /;
   ods output ScoreResults = Score;
   ods select FitPlot;
run;
```

Figure 12.14 Scatter Plot and Loess Model

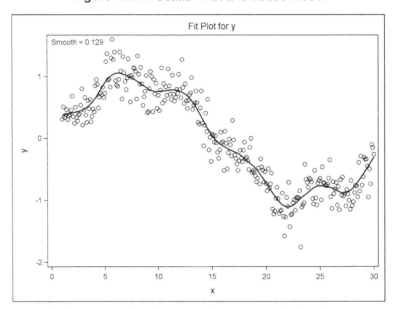

The Score data set contains the true signal, eta, and the predicted values, p_y. You can use the DATA step or PROC IML to compute the root mean square error, as follows:

```
proc iml;
use Score; read all var {f p_y}; close;
RMSE = sqrt( ssq(f-p_y)/nrow(f) );          /* SSQ = sum of squares */
print RMSE;
```

Figure 12.15 Root Mean Square Error

You can also use PROC SGPLOT to compare the true response and the predicted model:

```
proc sgplot data=Score;
   series x=x y=f / legendlabel="True Model";
   series x=x y=p_y / legendlabel="Loess(0.129)";
run;
```

Figure 12.16 True Response and Loess Model

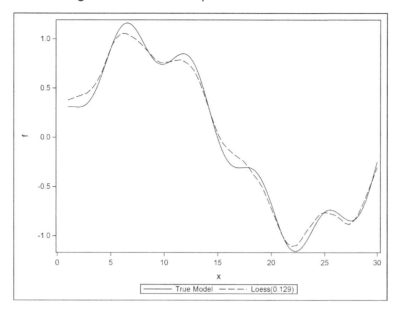

Nonparametric methods such as the LOESS procedure often involve tuning a parameter that tries to balance underfitting and overfitting the data. Often the smoothing parameter is chosen to minimize a criterion such as generalized cross validation or the Akaike information criterion. Sometimes different smoothing parameters result in models that reveal different aspects of the response. For example, the following exercise shows that you can also find a smoothing parameter that models the $\sin(x/5)$ component of the signal.

Exercise 12.8: Modify the previous analysis by using the SMOOTH=0.31 option in the MODEL statement to manually select a smoothing parameter value. Use PROC SGPLOT to compare the fitted smoother to the $\sin(x/5)$ component of the response variable.

12.6 References

Bender, R., Augustin, T., and Blettner, M. (2005), "Generating Survival Times to Simulate Cox Proportional Hazards Models," *Statistics in Medicine*, 24, 1713–1723.

Fan, X., Felsovályi, A., Sivo, S. A., and Keenan, S. C. (2002), *SAS for Monte Carlo Studies: A Guide for Quantitative Researchers*, Cary, NC: SAS Institute Inc.

Kleinman, K. and Horton, N. J. (2010), *Using SAS for Data Management, Statistical Analysis, and Graphics*, Boca Raton, FL: CRC Press.

Chapter 13
Simulating Data from Time Series Models

Contents

13.1 Overview of Simulating Data from Time Series Models

Many processes can be modeled with time series. Applications of time series modeling include forecasting demand for retail goods, predicting seasonal fluctuations of airline passengers, and managing risk in financial portfolios. Simulation plays a big part in time series modeling. It is used to analyze the sensitivity of a model and to ask "What if" questions such as "What if there is a recession?" or "What if there is a stock market crash?"

A characteristic of time series is that the observations are correlated with each other. When you want to simulate univariate time series data, you need to modify the simulation techniques presented in Chapter 2, "Simulating Data from Common Univariate Distributions." Whereas earlier chapters assumed that observation is drawn independently from a distribution, the techniques in this chapter simulate *autocorrelated* data. The autocorrelation of time series describes the correlation between values of the series at different time periods.

This chapter describes a few procedures and functions in SAS software that enable you to simulate data from time series. For univariate time series data, you can use either the DATA step or the ARMASIM function in SAS/IML software. To simulate multivariate time series data, use the SAS/IML VARMASIM function; it is unwieldy to use the DATA step for this purpose. Fan et al. (2002) present additional examples of simulating time series data with SAS software.

13.2   Simulating Data from Autoregressive and Moving Average (ARMA) Models

The regression models in Chapter 11, "Simulating Data for Basic Regression Models," are useful when you want to examine the distribution of parameter estimates from SAS/STAT regression procedures such as PROC GLM and PROC MIXED. In a similar way, you can use simulated time series data to examine the sampling distributions of statistics that are produced by SAS/ETS procedures such as PROC ARIMA and PROC MODEL.

Two simple time series models are the autoregressive (AR) and the moving average (MA) models.

13.2.1   A Simple Autoregressive Model

In an AR model, the data are a series of observations where the value at time t is linearly related to the values at one or more previous times, plus a random error term. If $\phi_1, \phi_2, \ldots, \phi_p$ are constants, then the AR model of order p (AR(p)) is as follows:

$$y_t = \Sigma_{i=1}^{p} \phi_i \, y_{t-i} + \epsilon_t$$

A particularly simple AR model assumes that the value at time t is related only to the value at the immediately preceding time. This is called an AR(1) model and has the form

$$y_t = \phi_1 y_{t-1} + \epsilon_t$$

As is the case for linear regression, the error terms ϵ_t are often chosen to be independent and to follow a normal distribution. The error terms are also called the *innovations*.

When simulating time series data, it suffices to simulate time series for which the series is expected to have zero mean and for which $\epsilon_t \sim N(0, 1)$. A simple linear transformation $Z = \sigma Y + \mu$ will transform the data to a series for which the expected mean is μ and for which $\epsilon_t \sim N(0, \sigma)$.

13.2.2   Simulating AR(1) Data in the DATA Step

When simulating data from a stationary time series, a common technique is to generate a few values that are not recorded so that the initial value of the time series is not constant. This is shown in the following DATA step, which throws away 11 points before the simulated data are saved:

```
%let N = 100;
data AR1(keep=t y);
call streaminit(12345);
phi = 0.4;    yLag = 0;
do t = -10 to &N;
   eps = rand("Normal");          /* variance of 1              */
   y = phi*yLag + eps;            /* expected value of Y is 0 */
   if t>0 then output;
   yLag = y;
end;
run;
```

You can visualize the time series and estimate the parameters by using the ARIMA procedure. The parameter estimates are shown in Figure 13.1.

```
ods graphics on;
proc arima data=AR1 plots(unpack only)=(series(corr));
   identify var=y nlag=1;                         /* estimate AR1 lag */
   estimate P=1;
   ods select SeriesPlot ParameterEstimates FitStatistics;
quit;
```

Figure 13.1 Parameter Estimates

The ARIMA Procedure

Conditional Least Squares Estimation					
Parameter	Estimate	Standard Error	t Value	Approx Pr > |t|	Lag
MU	-0.04315	0.13836	-0.31	0.7558	0
AR1,1	0.31518	0.09661	3.26	0.0015	1

Constant Estimate	-0.02955
Variance Estimate	0.90753
Std Error Estimate	0.952644
AIC	276.0646
SBC	281.2749
Number of Residuals	100

The series has three parameters. The expected value of the series is zero. The estimated value, MU, is shown in Figure 13.1. The estimate for MU is small and the corresponding p-value is large, which indicates that the data are consistent with the hypothesis that the parameter is zero.

The parameter estimate for ϕ is shown in the row with the AR1,1 heading. The parameter estimate is close to 0.4, which is the value assigned to the model.

The third parameter is the variance of the error term. The model uses a variance of 1, and the "Variance Estimate" row in Figure 13.1 shows that the estimated variance is close to its true value.

Figure 13.2 shows a plot of the time series. The plot indicates that the mean of the series is zero and that the standard deviation is approximately 1. Although is is difficult to estimate autocorrelation from this plot, you can see that an observation that is greater than zero (respectively, less than zero) tends to be followed by an observation that is also greater than zero (respectively, less than zero).

Exercise 13.1: In the DATA step, define $Z = 3Y + 2$. (Remember to add Z to the KEEP statement.) Run PROC ARIMA and specify VAR=Z in the IDENTIFY statement. How do the parameter estimates change?

Figure 13.2 Time Series for AR(1) Data

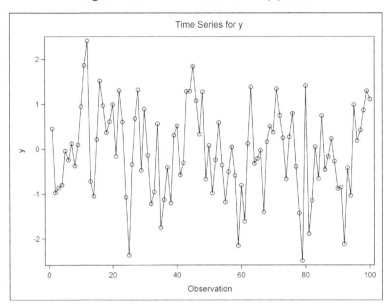

13.2.3   Approximating Sampling Distributions for the AR(1) Parameters

You can add a loop to the DATA step and use a BY statement in PROC ARIMA in order to approximate the sampling distributions of the statistics that are produced by PROC ARIMA. The PROC ARIMA statement does not support the NOPRINT option, but this option is supported in the IDENTIFY and ESTIMATE statements. The ESTIMATE statement also supports an OUTEST= option that creates an output data set. A WHERE clause is used to restrict the output to the three parameter estimates.

```
%let N = 100;
%let NumSamples = 1000;
/* simulate 1,000 time series from model */
data AR1Sim(keep=SampleID t y);
phi = 0.4;
call streaminit(12345);
do SampleID = 1 to &NumSamples;
   yLag = 0;
   do t = -10 to &N;
      y = phi*yLag + rand("Normal");
      if t>0 then output;
      yLag = y;
   end;
end;
run;

/* estimate AR(1) model for each simulated time series */
proc arima data=AR1Sim plots=none;
   by SampleID;
   identify var=y nlag=1 noprint;                    /* estimate AR1 lag */
   estimate P=1 outest=AR1Est(where=(_TYPE_="EST")) noprint;
quit;
```

There are several ways to analyze the sampling distributions. The following statements use PROC UNIVARIATE to create a histogram (see Figure 13.3) of the approximate sampling distribution for the estimate of the ϕ parameter. The sampling distribution is asymptotically normal for large samples, so a normal curve is overlaid on the histogram. The MEANS procedure displays descriptive statistics, which are shown in Figure 13.4.

```
/* analyze sampling distribution */
ods graphics on;
proc univariate data=AR1Est;
   var AR1_1;
   histogram AR1_1 / normal;
   ods select histogram;
run;

proc means data=Ar1Est Mean Std P5 P95;
   var AR1_1;
run;
```

Figure 13.3 Approximate Sampling Distribution of AR(1) Parameter Estimate

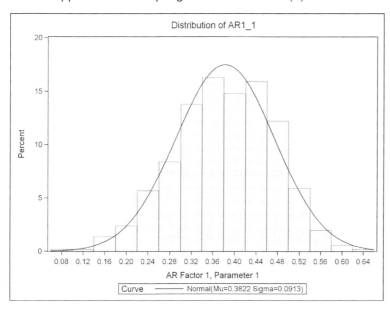

Figure 13.4 Descriptive Statistics of AR(1) Sampling Distribution

The MEANS Procedure

Analysis Variable : AR1_1 AR Factor 1, Parameter 1			
Mean	Std Dev	5th Pctl	95th Pctl
0.3822170	0.0912675	0.2305033	0.5203583

The standard deviation of the sampling distribution is the standard error of the parameter estimate. The Monte Carlo estimate in Figure 13.4 is similar to the estimate in Figure 13.1. However, the

Monte Carlo standard error in Figure 13.4 is a better estimate than the point estimate in Figure 13.1, which was computed for a single time series with 100 observations.

Exercise 13.2: Visualize the sampling distribution of the variance of the error term and of the mean of the AR(1) process. Estimates for these quantities are in the ErrorVar and MU variables, respectively, in the AR1Est data set.

13.2.4   Simulating AR and MA Data in the DATA Step

In addition to the AR models, you can also simulate data from an MA model, or from a so-called ARMA model that combines the characteristics of the AR and MA models.

An MA model assumes that the error term at time t is linearly related to error terms at previous times. If $\theta_1, \theta_2, \ldots, \theta_q$ are constants, then the MA model of order q (MA(q)) is as follows:

$$y_t = \epsilon_t - \Sigma_{i=1}^{q} \theta_i \epsilon_{t-i}$$

The *SAS/ETS User's Guide* cautions that "Moving-average errors can be difficult to estimate." The documentation further notes that "A moving-average process can usually be well-approximated by an autoregressive process."

The AR and MA models can be combined into a model that includes both kinds of terms. A general ARMA(p, q) model has the following form:

$$y_t = \Sigma_{i=1}^{p} \phi_i y_{t-i} + \epsilon_t - \Sigma_{i=1}^{q} \theta_i \epsilon_{t-i}$$

Again, the *SAS/ETS User's Guide* cautions that "ARMA models can be difficult to estimate," and discusses issues that lead to "instability of the parameter estimates."

Although checking the parameter estimates might be problematic, you can simulate data from an ARMA(p, q) model by using the DATA step. The following DATA step incorporates lagged effects by keeping previous values in an array. Although the program assumes that $p > 0$ and $q > 0$, you can easily modify the program to simulate, for example, data from a pure MA model. (Or simply set the AR coefficients to zero.)

```
/* simulate data from ARMA(p,q) model */
%let N = 5000;
%let p = 3;                                    /* order of AR terms */
%let q = 1;                                    /* order of MA terms */
data ARMASim(keep= t y);
call streaminit(12345);
array phi phi1 - phi&p (-0.6, 0, -0.3);        /* AR coefficients    */
array theta theta1 - theta&q ( 0.1);           /* MA coefficients    */
array yLag   yLag1 - yLag&p;          /* save p lagged values of y */
array errLag errLag1 - errLag&q;      /* save q lagged error terms */
```

```
/* set initial values to zero */
do j = 1 to dim(yLag);   yLag[j] = 0;   end;
do j = 1 to dim(errLag); errLag[j] = 0; end;

/* "steady state" method: discard first 100 terms */
do t = -100 to &N;
   /* y_t is a function of values and errors at previous times      */
   e = rand("Normal");
   y = e;
   do j = 1 to dim(phi);                                    /* AR terms */
      y = y + phi[j] * yLag[j];
   end;
   do j = 1 to dim(theta);                                  /* MA terms */
      y = y - theta[j] * errLag[j];
   end;
   if t > 0 then output;

   /* update arrays of lagged values */
   do j = dim(yLag) to 2 by -1; yLag[j] = yLag[j-1]; end;
   yLag[1] = y;
   do j = dim(errLag) to 2 by -1; errLag[j] = errLag[j-1]; end;
   errLag[1] = e;
end;
run;
```

The program is organized as follows:

1. The arrays **phi** and **theta** are used to hold the AR and MA coefficients, respectively.

2. The arrays **yLag** and **errLag** are used to hold the lagged values of the response and the error terms, respectively.

3. Zeros are assigned as initial values for all lagged terms. This is somewhat arbitrary, but it can be defended by noting that the series is assumed to have mean value 0 and the error terms also have mean value 0. Therefore, mean values are used in place of the unknown past values. However, notice that the p lagged values of Y do not have the correct autocovariance.

4. To make up for the fact that the initial values of Y do not have the correct autocovariance, the simulation discards the Y values for some initial period. In the program, "time" starts at $t = -100$, but data are not written to the data set until $t > 0$. By this time, the series should have settled into a steady state for which the response and error terms have the correct covariance.

5. The response at time t is computed according to the ARMA(p, q) model.

6. The arrays of lagged terms are updated.

You can use the SGPLOT procedure to visualize the stationary time series. The model depends on lags of order 1 and 3, so Figure 13.5 shows tick marks every 12 time periods.

```
proc sgplot data=ARMASim(obs=204);
   series x=t y=y;
   refline 0 / axis=y;
   xaxis values=(0 to 204 by 12);
run;
```

Figure 13.5 Simulated ARMA(3,1) Data

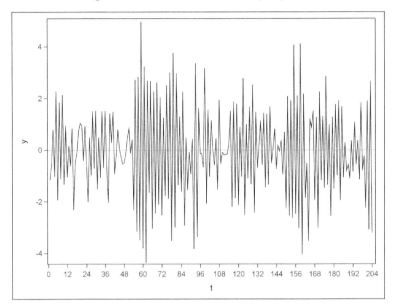

Exercise 13.3: Use PROC ARIMA analysis to estimate the AR and MA parameters. Because the series is so long ($N = 5000$), the parameter estimates are close to the model parameters. However, for short ARMA(p, q) time series this might not be the case.

Exercise 13.4: Rerun the simulation for $N = 100, 500, 1000,$ and 2000 and use PROC ARIMA to estimate the parameters, as was done in Section 13.2.3. Use PROC SGPLOT to plot the parameter estimates and standard errors as a function of the sample size.

13.2.5   Simulating AR(1) Data in SAS/IML Software

You can use the DATA step to simulate data from ARMA models, but it is much easier to use the ARMASIM function, which is part of SAS/IML software. The ARMASIM function simulates data from an AR, MA, or ARMA model.

The ARMASIM function uses a parameterization for the ARMA model in which the AR coefficients differ in sign from the formula in the previous section. The model used by the ARMASIM function has the following form:

$$\Sigma_{i=0}^{p} \phi_i \, y_{t-i} = \Sigma_{i=0}^{q} \theta_i \epsilon_{t-i}$$

where $\phi_0 = \theta_0 = 1$. There are other SAS/IML functions (such as ARMACOV and ARMALIK) that use the same parameterization.

An AR(1) time series with the same parameters as in Section 13.2.2 can be constructed as follows:

```
/* use the ARMASIM function to simulate a time series */
%let N = 100;
proc iml;
phi   = {1 -0.4};     /* AR coefficients: Notice the negative sign!   */
theta = {1};          /* MA coefficients: Use {1, 0.1} to add MA term */
mu = 0;               /* mean of process                             */
sigma = 1;            /* std dev of process                          */
seed = -54321;        /* use negative seed if you call ARMASIM twice  */
yt = armasim(phi, theta, mu, sigma, &N, seed);
```

Notice that the second element in the `phi` vector has the opposite sign as the corresponding coefficient in the DATA step in Section 13.2.2. Also notice that the first elements in the `phi` and `theta` vectors are always 1 because they represent the coefficients of the y_t and ϵ_t terms.

The ARMASIM function does not use the random number seed that is set by the RANDSEED subroutine (see Section 3.3). Instead, it maintains its own seed value. If you call ARMASIM twice with the same positive seed, then you will get exactly the same time series. In contrast, if you call ARMASIM twice with the same negative seed, then the second time series is different from the first, as shown by the following statements. The results are shown in Figure 13.6.

```
y1 = armasim(phi, theta, mu, sigma, 5,  12345);  /* 5 obs           */
y2 = armasim(phi, theta, mu, sigma, 5,  12345);  /* same 5 obs      */
y3 = armasim(phi, theta, mu, sigma, 5, -12345);  /* 5 obs           */
y4 = armasim(phi, theta, mu, sigma, 5, -12345);  /* different 5 obs */
print y1 y2 y3 y4;
```

Figure 13.6 Time Series with the Same Positive and Negative Seeds

y1	y2	y3	y4
-0.046896	-0.046896	0.5448519	-1.575105
-0.118747	-0.118747	-1.303247	-0.181169
-0.290992	-0.290992	0.2705028	-0.716656
-0.338658	-0.338658	0.6804066	1.4124796
-0.06193	-0.06193	0.4478731	0.5964082

In Figure 13.6, the vectors `y1` and `y2` have the exact same values, which is not usually very useful. However, the vectors `y3` and `y4`, which use a negative seed value, have different values. Consequently, it is best to use a negative seed when you simulate an ARMA model.

You can use PROC IML to simulate multiple ARMA samples, as was done in Section 13.2.3. The following SAS/IML program generates the data and also creates a `SampleID` variable:

```
NumSamples = 1000;
Y = j(NumSamples, &N);
do i = 1 to NumSamples;
   z = armasim(phi, theta, mu, sigma, &N, -12345);  /* use seed < 0 */
   Y[i,] = z`;                           /* put i_th time series in i_th row */
end;
SampleID = repeat(T(1:NumSamples),1,&N);
create AR1Sim var {SampleID Y}; append; close AR1Sim;
quit;
```

The program demonstrates a useful technique. Each simulated time series is stored in a row of the Y matrix. The **SampleID** matrix is a matrix of the same dimensions that has the value i for every element of the ith row. Recall that SAS/IML matrices are stored in row-major order. Therefore, when you use the CREATE statement and specify the **SampleID** and **Y** matrices, the elements of those matrices are written (in row-major order) to two long variables in the data set. There is no need to use the SHAPE function to create column vectors prior to writing the data.

Exercise 13.5: Create a SAS data set that contains the **yt** vector. Call PROC ARIMA and compare the parameter estimates to the estimates in Figure 13.1.

Exercise 13.6: Re-create Figure 13.3 and Figure 13.4 by using PROC ARIMA to analyze the Y variable in the AR1Sim data set. Mimic the PROC ARIMA code that is shown in Section 13.2.3.

13.3　Simulating Data from Multivariate ARMA Models

Fan et al. (2002) show examples of using the DATA step to simulate data from multivariate correlated time series where each component is simulated from an ARMA model. A multivariate time series of this type is known as a VARMA model.

The VARMA model includes matrices of parameters and vectors of response variables. Accordingly, simulating these data by using the DATA step requires writing out the matrix multiplication by hand. Fortunately, the SAS/IML language supports the VARMASIM subroutine, which has a syntax that is similar to the ARMASIM function.

For simplicity, consider a multivariate stationary VARMA(1,1) model with mean vector **0**. If there are k components, then the VARMA(1,1) model is

$$y_t = \Phi y_{t-1} + \epsilon_t - \Theta \epsilon_{t-1}$$

The $k \times k$ matrix Φ contains the AR coefficients. The (i, j)th element of Φ is the coefficient that shows how the ith component of y_t depends on the jth component of y_{t-1}. The vector ϵ_t (the vector of innovations) is assumed to be multivariate normal with mean vector **0** and $k \times k$ covariance matrix Σ. The $k \times k$ matrix Θ contains the MA coefficients.

The following SAS/IML program simulates a trivariate ($k = 3$) stationary time series that has no MA coefficients ($\Theta = 0$). The errors have covariance given by the matrix Σ, and the AR coefficients are given by the **Phi** matrix.

```
proc iml;
Phi = {0.70 0.00 0.00,            /* AR coefficients      */
       0.30 0.60 0.00,
       0.10 0.20 0.50};
Theta = j(3,3,0);                 /* MA coefficients = 0  */
mu = {0,0,0};                     /* mean = 0             */
sigma = {1.0  0.4 0.16,           /* covariance of errors */
         0.4  1.0 0.4,
         0.16 0.4 1.0};
```

```
call varmasim(y, Phi, Theta, mu, sigma, &N) seed=54321;

create MVAR1 from y[colname={"y1" "y2" "y3"}];
append from y;
close MVAR1;
quit;
```

To verify that the simulated data are from the specified model, you can use the STATESPACE procedure to estimate the parameters and covariances. Figure 13.7 shows the parameter estimates for the AR coefficients and the covariance of the error terms (innovations) based on 100 simulated observations. The estimates of the AR coefficients are contained in the "Estimates of Transition Matrix" table. The estimates are somewhat close to the model parameters, except for the Φ_{13} parameter. The covariance estimates are also reasonably close but, as you might guess, the standard errors for these statistics are large when the time series contains only 100 observations.

```
proc statespace data=MVAR1 interval=day armax=1 lagmax=1;
   var y1-y3;
   ods select FittedModel.TransitionMatrix FittedModel.VarInnov;
run;
```

Figure 13.7 Multivariate AR(1) Time Series with Correlated Innovations

The STATESPACE Procedure
Selected Statespace Form and Fitted Model

Estimate of Transition Matrix		
0.688963	0.028057	-0.12454
0.359162	0.489038	-0.02267
0.150481	0.117534	0.636814

Variance Matrix for Innovation		
0.907677	0.355207	0.14475
0.355207	0.866594	0.358142
0.14475	0.358142	0.834335

Exercise 13.7: Simulate 1,000 data samples from the VARMA model. Analyze the results with PROC STATESPACE and compute estimates for the standard errors of the AR coefficients and the elements of the innovation covariance.

13.4   References

Fan, X., Felsovályi, A., Sivo, S. A., and Keenan, S. C. (2002), *SAS for Monte Carlo Studies: A Guide for Quantitative Researchers*, Cary, NC: SAS Institute Inc.

Chapter 14
Simulating Data from Spatial Models

Contents

14.1 Overview of Simulating Data from Spatial Models

Chapter 13, "Simulating Data from Time Series Models," describes how to simulate observations that are correlated in time, which is a one-dimensional quantity. Spatial models provide an example of two-dimensional correlated observations. For spatial data, the correlations are typically positive.

You can use spatial simulation to understand the uncertainty in models that interpolate spatial data (Kolovos 2010). For two-dimensional spatial models, a primary simulation tool is the SIM2D procedure in SAS/STAT software. The SIM2D procedure can simulate a Gaussian random field with a specified mean and a variety of standard covariance structures. This chapter describes how to simulate conditional and unconditional random fields. The chapter also presents algorithms and SAS/IML programs for simulating data from spatial point processes. For additional examples of simulating spatially correlated data, see Schabenberger and Gotway (2005).

14.2 Simulating Data from a Gaussian Random Field

In SAS software, you can use the SIM2D procedure to simulate data from a two-dimensional Gaussian random field. The documentation for the SIM2D procedure in the *SAS/STAT User's Guide* describes the Gaussian random field model in detail. Briefly, let $s = (x, y)$ be an arbitrary point in some region of the plane, and let the points $s_1, s_2, \ldots, s_k$ be locations that you specify.

The SIM2D procedure enables you to produce random vectors $\mathbf{Z} = (Z(s_1), Z(s_2), \ldots, Z(s_k))^T$ with a specified covariance structure and mean. The field is *Gaussian* because it can be written as $\mathbf{Z}(s) = \mu(s) + \epsilon(s)$, where $\mu(s)$ is the mean function and $\epsilon(s)$ is multivariate normal with zero mean and a specified covariance.

In practice, the covariance matrix is often assumed to be generated by a spatial covariance function, V, which is a function of the distance, h, between two points. An example of a spatial covariance function is the Gaussian isotropic covariance structure $V(h) = c_0 \exp(-h^2/a_0^2)$, where h is the Euclidean distance between two points. The parameters c_0 and a_0 are called the *sill* (or scale) and the *range*, respectively. For details about various forms of covariance structures, see Schabenberger and Gotway (2005) or the documentation for the VARIOGRAM procedure in the *SAS/STAT User's Guide*.

The SIM2D procedure performs two kinds of simulation. The documentation focuses on *conditional simulation* in which you provide a data set with (x, y) coordinates and measurements (z) at those coordinates. Given a mean function and a covariance structure, the SIM2D procedure can produce simulations that "honor the data," which means that the simulated values at the observed locations match the observed values (Schabenberger and Gotway 2005, p. 406).

You can also use PROC SIM2D to produce *unconditional simulations*. In an unconditional simulation, no observed data are used to constrain the values in the simulation. Both simulation methods are described in the "Details" section of the PROC SIM2D documentation.

14.2.1   Unconditional Simulation of One-Dimensional Data

An unconditional simulation is a model of the form $Z(s) = \mu(s) + \epsilon(s)$, where $\mu(s)$ is the mean function and $\epsilon(s)$ is multivariate normal with zero mean and covariance matrix

$$
C_z = \begin{pmatrix}
C(0) & C(s_1 - s_2) & \cdots & C(s_1 - s_k) \\
C(s_2 - s_1) & C(0) & \cdots & C(s_2 - s_k) \\
& & \ddots & \\
C(s_k - s_1) & C(s_k - s_2) & \cdots & C(0)
\end{pmatrix}
$$

To illustrate unconditional simulation, consider the case of evenly spaced locations in one dimension. Assume that the covariance function is the Gaussian function $C(h) = c_0 \exp(-h^2/a_0^2)$. The following SAS/IML program computes the covariance function for evenly spaced points in the interval $[0, 100]$. The result is shown in Figure 14.1.

```
proc iml;
/* Define Gaussian covariance function V(s). Each row of s is a point.
   c0 = scale;  a0=range;  s is n x p matrix */
start GaussianCov(c0, a0, s);
   h = distance(s);            /* n x n matrix of pairwise distances  */
   return ( c0#exp( -(h##2/a0##2) ) );
finish;


/* 1D example of unconditional distrib of spatially correlated data */
scale = 8;   range = 30;   /* params for the Gaussian random field */
s = do(0, 100, 25);        /* evaluate field at these locations    */
Cov = GaussianCov(scale, range, s`);
print Cov[format=6.4 r=("s1":"s5") c=("s1":"s5")];
```

Figure 14.1 Spatial Covariance Matrix

			Cov		
	s1	**s2**	**s3**	**s4**	**s5**
s1	8.0000	3.9948	0.4974	0.0154	0.0001
s2	3.9948	8.0000	3.9948	0.4974	0.0154
s3	0.4974	3.9948	8.0000	3.9948	0.4974
s4	0.0154	0.4974	3.9948	8.0000	3.9948
s5	0.0001	0.0154	0.4974	3.9948	8.0000

The computation of the Gaussian covariance function uses the DISTANCE function, which was introduced in SAS/IML 12.1. If you are running an earlier version, then use the EUCLIDEANDISTANCE function that is provided in Appendix A, "A SAS/IML Primer."

Figure 14.1 shows that the variance of the field at each point is 8. The covariance of the field between neighboring points, which are 25 units apart, is about 4. Between points that are separated by 50 units, the covariance is about 0.5. The field values at points that are 100 units apart are essentially independent.

You can visualize realizations of the field by using the RANDNORMAL function, which is described in Section 8.3.1. The following function generates realizations of the Gaussian random field at specified locations. The function is called to generate four realizations of a field at the points $s_1 = 0, s_2 = 25, \ldots, s_5 = 100$. Figure 14.2 shows four realizations of the field.

```
/* Simulate GRF unconditionally. Each row of s is a point.
   s is n x p matrix of n points in p-dimensional space */
start UncondSimGRF(NumSamples, s, param);
   mean = param[1]; scale=param[2]; range=param[3];
   C = GaussianCov(scale, range, s);
   return( RandNormal( NumSamples, j(1,nrow(s),mean), C) );
finish;

call randseed(12345);
mean = 40;
Z = UncondSimGRF(4, s`, mean||scale||range);       /* 4 realizations */
print (s`)[label="Location"] (Z`)[format=5.3 c=("z1":"z4")];
```

Figure 14.2 Spatial Covariance Matrix

Location	z1	z2	z3	z4
0	40.75	36.51	39.58	40.77
25	43.01	37.91	40.50	39.84
50	43.63	42.04	37.69	40.22
75	40.19	41.74	37.13	39.31
100	42.93	43.20	37.95	39.97

Notice that the values in Figure 14.2 are centered around 40, which is the mean value of the field. You can plot several realizations on a finer grid to see how the process varies at each point. The following statements generate six realizations of the field at evenly spaced points that are 5 units apart. These realizations are shown in Figure 14.3.

```
s = do(0, 100, 5);        /* evaluate the field at these locations */
NumSamples = 6;
Z = UncondSimGRF(NumSamples, s`, mean||scale||range);

/* Each row of z is a realization. Construct X and SampleID vars    */
N = ncol(s);
SampleID = repeat( T(1:NumSamples), 1, N );
x = repeat( s, 1, NumSamples );
create GRF1D var {"SampleID" "x" "z"};  append;  close GRF1D;
quit;

proc sgplot data=GRF1D;
   title "Realizations of a Gaussian Random Field";
   series x=x y=z / group=SampleID;
   refline 40 / axis=y;
run;
```

Figure 14.3 Six Realizations of a Gaussian Random Field

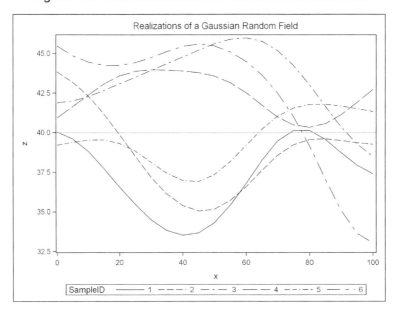

Figure 14.3 shows that each realization looks like a smooth function whose mean value is close to 40. The values of the field are generated from a multivariate normal distribution with covariance matrix C_Z.

If you simulate many realizations, then you can examine the variation about the mean of the process at any location. For example, suppose that GRF is the name of a data set that contains 1,000 realizations. The following statements use PROC MEANS to show the variation at three spatial locations. See Figure 14.4.

```
proc means data=GRF Mean Var Min Max;
   where x in (20 50 80);
   class x;
   var z;
run;
```

Figure 14.4 Variation at Three Spatial Positions

The MEANS Procedure

		Analysis Variable : z			
x	N Obs	Mean	Variance	Minimum	Maximum
20	1000	39.8186998	7.9781623	29.7776349	47.7638947
50	1000	39.8894179	8.2698529	27.9176718	50.9783990
80	1000	40.1238658	7.9379840	31.1521781	50.3380485

14.2.2   Unconditional Simulation of Two-Dimensional Data

The previous section uses PROC IML to simulate one-dimensional spatial data. You can also use PROC IML to simulate Gaussian random fields in higher dimensions. However, for the important case of two-dimensional data use the SIM2D procedure, which contains many features that are useful for simulations.

Suppose that you want to unconditionally simulate from a two-dimensional random field. The following statements define the same Gaussian covariance function that was used in the previous section, but uses it to simulate four realizations of a field on a grid of points in the region $[0, 100] \times [0, 100]$.

```
/* unconditional simulation by using PROC SIM2D */
proc sim2d outsim=GRF narrow;
   grid x = 0 to 100 by 10
        y = 0 to 100 by 10;
   simulate form=Gauss scale=8 range=30 numreal=4 seed=12345;
   mean 40;
run;
```

The procedure has four statements:

- The PROC SIM2D statement specifies that the realizations should be stored in the GRF data set. The NARROW option specifies that certain nonessential variables are not output.

- The GRID statement specifies that each simulated field should be evaluated on a regular grid defined by the points (x, y), where x and y are in the set $\{0, 10, \ldots, 100\}$.

- The SIMULATE statement defines the covariance function. The FORM= option specifies that the covariance function is the Gaussian function $C(h) = c_0 \exp(-h^2/a_0^2)$, and the SCALE= and RANGE= options specify the c_0 and a_0 parameters, respectively. The NUMREAL= option specifies that the procedure should simulate four realizations.

- The MEAN statement specifies that the field has a mean function that is a constant.

Each statement has other options that are not shown. For example, you can use the PLOTS= option in the PROC SIM2D statement to construct a graph that helps you to summarize the simulated data. The FORM= option can specify several covariance functions, which are defined in the "Details" section of the PROC VARIOGRAM documentation in the *SAS/STAT User's Guide*. The MEAN statement can be used to specify mean functions that are quadratic functions of the coordinates.

The OUTSIM= data set contains four variables. The _ITER_ variable identifies the data that are generated for each simulation. The GXC and GYC variables contain the horizontal and vertical coordinates of the specified locations. The SValue variable contains the values of the realization at each location. You can use the following Graph Template Language (GTL) template to define a template for a contour plot. The contour plot, which is shown in Figure 14.5, visualizes a single realization:

```
/* define a contour plot template */
proc template;
define statgraph ContourPlotParm;
dynamic _X _Y _Z _TITLE;
begingraph;
   entrytitle _TITLE;
   layout overlay;
      contourplotparm x=_X y=_Y z=_Z / nhint=12
        contourtype=fill colormodel=twocolorramp name="Contour";
      continuouslegend "Contour" / title=_Z;
   endlayout;
endgraph;
end;
run;


proc sgrender data=GRF template=ContourPlotParm;
where _ITER_ = 1;                               /* or use BY _ITER_ */
dynamic _TITLE="Realization of a 2D Gaussian Random Field"
        _X="gxc" _Y="gyc" _Z="SValue";
run;
```

Figure 14.5 One Realization of a Two-Dimensional Gaussian Random Field

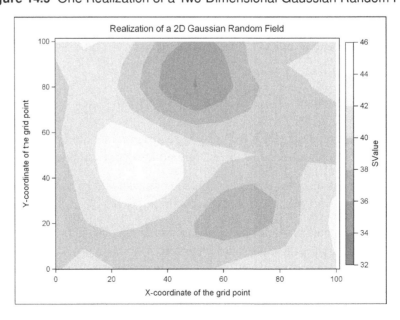

Figure 14.5 shows that the contours of the realized surface are close to the mean value of 40. This fact is true for all of the realized surfaces, but the locations of the surface optima vary from surface to surface, just as in Figure 14.3, which shows the one-dimensional case.

The SIM2D procedure can produce a contour plot that summarizes the simulated realizations. The following statements generate 5,000 realizations and request the "SimulationPlot" plot, which shows contours of the mean values of the realizations and uses color to indicate standard deviations. From the scale of the contours and the color ramp, you can argue that the underlying field is essentially constant with a constant variance $\sqrt{8} \approx 2.83$. The contour plot is shown in Figure 14.6.

```
/* unconditional simulation */
proc sim2d outsim=GRF plots=(sim);
   grid x = 0 to 100 by 10
        y = 0 to 100 by 10;
   simulate scale=8 range=30 form=Gauss numreal=5000 seed=12345;
   mean 40;
run;
```

Figure 14.6 Summary of Simulations

The color ramp in Figure 14.6 uses white to represent low values and a dark color to represent high values. However, this assignment of colors might vary according to the ODS style that you are using to generate the ODS graphics. For example, the Journal style uses dark colors to represent low values, whereas the HTMLBlue style uses dark colors to represent high values.

Exercise 14.1: Create contour plots of the other three surfaces in the GRF data set.

Exercise 14.2: Use the UNCONDSIMGRF function in SAS/IML (see Section 14.2.1) to produce realizations of the two-dimensional Gaussian random field on a coarse grid of locations.

Exercise 14.3: At a specified location such as $(50, 50)$, the Gaussian random field is normally distributed. Rerun the PROC SIM2D program. Use PROC UNIVARIATE to display a histogram and to fit a normal distribution to the simulated data at $(50, 50)$.

14.2.3   Conditional Simulation of One-Dimensional Data

Conditional simulation means that the value of the field at certain locations is known. The canonical situation is that you have observed values at certain locations and want to simulate values at other locations. The simulated values at the unobserved locations are conditioned on the observed values. Because of the spatial correlation structure, the value of the field at an unobserved location, $Z(s)$, is expected to be close to observed values for locations near s.

Conditional simulation of multivariate normal data is described in Section 8.6 where the CONDMVN function is used to conditionally simulate data from a p-dimensional multivariate normal distribution. The CONDMVN function conditions on the values of the last k variables. In spatial statistics, the locations of the observed data might be interspersed among the unobserved locations. In order to reuse the CONDMV function, it is useful to construct two simple helper functions. The first function rearranges columns in a matrix so that the conditioning variables are the last k variables. The second function restores the columns to their original order for analysis and visualization. The following SAS/IML statements define these helper functions:

```
proc iml;
load module=(GaussianCov CondMVN);              /* load from storage */

/* Move the k columns specified by idx to the end of y */
start MoveColsToEnd(y, idx);
   i = setdif(1:ncol(y), idx);
   return( y[ , i||idx] );
finish;

/* Move last k cols of y to positions specified by idx, k=ncol(idx) */
start ReorderLastCols(y, idx);
   p = ncol(y);    k = p - ncol(idx);
   i = setdif(1:p, idx);                  /* indices of other columns    */
   v = j(nrow(y),p);                      /* result vector               */
   v[,idx] = y[ ,k+1:p];                  /* last k columns move to idx  */
   v[,i] = y[ ,1:k];                      /* others interspersed         */
   return( v );
finish;
```

With these helper functions defined, the following steps simulate conditional realizations of a Gaussian random field for one-dimensional data:

1. Start with observed data $(x_1, y_1), (x_2, y_2), \ldots, (x_k, y_k)$, where the y_i are the observed values of the field at the locations x_i.

2. Define the vector $v = (s_1, s_2, \ldots, s_{n-k}) \cup (x_1, \ldots, x_k)$, where the s_i are the locations at which you want to simulate data, given the x_i.

3. Compute the spatial covariance matrix, C, for the locations in v.

4. Call the CONDMVN function to simulate values z_i for the random field $Z(s_i)$, $i = 1, \ldots, n - k$.

5. If desired, rearrange the order of the vector $w = (z_1, z_2, \ldots, z_{n-k}) \cup (y_1, \ldots, y_k)$ to visualize the results.

This scheme is implemented by the following statements:

```
/* conditional simulation */
scale=8;  range=30;  mean=40;             /* params for GRF           */
s = do(0, 100, 5);
n = ncol(s);
idx = {3 10 19};                          /* s[idx] are observed locs */
y  = {41 39.25 39.75};                    /* observed values          */

v = MoveColsToEnd(s,idx);                 /* put observed locs last    */
C = GaussianCov(scale, range, v`);        /* compute covariance matrix */
z = CondMVN(1, j(1,n,mean), C, y);        /* conditional simulation    */
w = ReorderLastCols(z||y,idx);            /* restore to spatial order  */
```

If you plot the vector **w** against the vector **s**, then you will see one realization of the conditional
simulation. The following statements generate six realizations and visualize them by using PROC
SGPLOT. Figure 14.7 shows that the conditional realizations all pass through the observed values at
the corresponding locations. The curves are "pinched" at the locations for which observed data exist.
The field is constrained to exhibit less variation near the observed locations.

```
NumSamples=6;
z = CondMVN(NumSamples, j(1,n,mean), C, y);
w = ReorderLastCols(z||repeat(y,NumSamples),idx);
SampleID = repeat( T(1:NumSamples), 1, N );   /* create ID variable */
x = repeat(s,1,NumSamples);
create UGRF1D var {"SampleID" "x" "w"};  append;  close UGRF1D;
quit;

proc sgplot data=UGRF1D;
   title "Conditional Realizations of a Gaussian Random Field";
   series x=x y=w / group=SampleID;
   refline 40 / axis=y;
run;
```

Figure 14.7 Six Conditional Realizations of a Gaussian Random Field

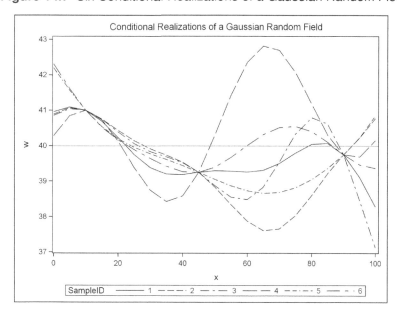

14.2.4 Conditional Simulation of Two-Dimensional Data

The documentation for the SIM2D procedure contains several examples of simulation conditioned on real data in the Sashelp.Thick data set, which contains measurements of the thickness of a coal seam at various spatial locations. The data set contains three variables: the East and North variables contain the coordinates of 121 observed locations, and the Thick variable contains the measured thickness at those same locations.

PROC VARIOGRAM is used to model the spatial correlation of the coal seam data. The analysis concludes that "a reasonable approximation of the spatial process generating the coal seam data is given by $Z(s) = \mu + \epsilon(s)$ where $\epsilon(s)$ is a Gaussian [random field] with Gaussian covariance structure $C_Z(h) = c_0 \exp(-h^2/a_0^2)$." The VARIOGRAM procedure estimates the parameter values. The parameter estimates are close to the values that are used in the previous sections: $\mu = 40, c_0 = 8,$ and $a_0 = 30$.

Implementing a two-dimensional conditional simulation is no more difficult than the unconditional simulation in Section 14.2.2. There are three changes that you need to make to the PROC SIM2D statement that created Figure 14.6:

1. Add **DATA=Sashelp.Thick** to the PROC SIM2D statement. This option specifies that the procedure should use this data to condition the simulations.

2. Add a COORDINATES statement. The COORDINATES statement specifies the variables in the data set that contain the observed spatial locations.

3. Add **VAR=Thick** to the SIMULATE statement. This option specifies that the observed values are contained in the Thick variable in the Sashelp.Thick data set.

The following call to PROC SIM2D adds these three items and creates Figure 14.8.

```
/* conditional simulation */
proc sim2d data=Sashelp.Thick outsim=GRF plots=(sim);
   coordinates xc=East yc=North;
   grid x = 0 to 100 by 10
        y = 0 to 100 by 10;
   simulate var=Thick
           scale=8 range=30 form=Gauss numreal=5000 seed=12345;
   mean 40;
run;
```

Figure 14.8 shows a summary plot of 5,000 conditional simulations. The mean of the simulated fields reflects the values of the observed measurements. The standard deviations at most points are small because the 121 observed locations were scattered throughout the regions. The dark areas in the corners and along the edges are locations that are far away from the data. Consequently, the realizations at those locations have a greater standard deviation than at locations that are close to observed values.

Exercise 14.4: Use the PROC SGRENDER statement that generates Figure 14.5 to visualize a few realizations of the conditional simulation.

Figure 14.8 Summary of Conditional Simulations

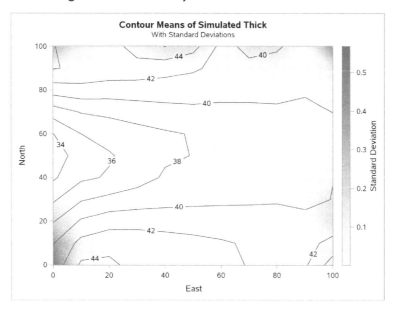

14.3   Simulating Data from a Spatial Point Process

You can use techniques from previous chapters to generate data from point processes with specified characteristics such as intensity, regularity, or clustering.

This section describes the simulation of the homogeneous and inhomogeneous Poisson point processes in two-dimensional spatial regions. A *Poisson process* generates points in a region. The density of the points is determined by the *intensity* of the process. (Here "intensity" means first-order intensity.) If the point process has constant intensity across the spatial region, then it is said to be *homogeneous*. If the intensity varies, then it is said to be *inhomogeneous*.

The definitions of these quantities are easiest if the planar region is rectangular. (Section 14.4 describes how to handle simulations on nonrectangular domains.) Let $A = [0, a] \times [0, b]$ be a rectangle. A homogeneous Poisson process on A satisfies the following:

- The number of points in A (sometimes denoted $N(A)$) is a random variable that follows a Poisson distribution with mean $\lambda |A|$, where $|A| = ab$ is the area of the region. The value λ is the intensity of the process.

- For each realization, the points in A are a random sample drawn independently from the uniform distribution on A.

A homogeneous Poisson process is devoid of spatial structure. Such a pattern is said to exhibit *complete spatial randomness*.

The definition of an inhomogeneous Poisson process is similar, except that the intensity is allowed to be a spatial function, $\lambda(x, y)$. For an inhomogeneous process, the first condition is replaced by the following:

- $N(A)$ is a random variable that follows a Poisson distribution with mean $\int_A \lambda(x, y) \, dx \, dy$. On the rectangle, this simplifies to $\int_0^b \int_0^a \lambda(x, y) \, dx \, dy$.

14.4   Simulating Data from a Homogeneous Poisson Process

The definition of a homogeneous Poisson process requires sampling n points from the uniform distribution on the rectangle $A = [0, a] \times [0, b]$. This is easily done in the SAS/IML language by passing an $n \times 2$ matrix to the RANDGEN subroutine, as shown in the following function definition:

```
proc iml;
/* simulate n points uniformly and independently on [0,a]x[0,b] */
start Uniform2d(n, a, b);
   u = j(n, 2);
   call randgen(u, "Uniform");
   return( u # (a||b) );                    /* scale to [0,a]x[0,b] */
finish;
```

In the function, the matrix **u** is filled with uniform variates on the unit square. These variate are scaled onto the rectangle A by using elementwise multiplication: the first column is multiplied by a and the second column is multiplied by b.

With this helper function defined, you can define a function to generate a homogeneous Poisson process on A. The function takes the intensity of the process as an argument, as follows:

```
start HomogPoissonProcess(lambda, a, b);
   n = rand("Poisson", lambda*a*b);
   return( Uniform2d(n, a, b) );
finish;
```

To test the function, you can generate a realization on the rectangle $[0, 3] \times [0, 2]$. If you specify an intensity of $\lambda = 100/6$, then the expected number of points is 100.

```
call randseed(123);
a = 3; b = 2;
u = HomogPoissonProcess(100/(a*b), a, b);
```

Figure 14.9 shows the 94 points that are generated. The distribution of the points is consistent with a two-dimensional uniform distribution and the number of points is consistent with the claim that $N(A)$ is a Poisson random variable with mean value 100.

To simulate a point process in a nonrectangular region, R, generate points in a rectangle that contains R, and exclude any points that are not in R. The resulting points are a realization of a homogeneous Poisson process on R.

Exercise 14.5: Use the DATA step to implement the algorithm for simulating a homogeneous Poisson process.

Exercise 14.6: Simulate a homogeneous Poisson process on the triangle determined by the points $(0, 0)$, $(a, 0)$, and $(0, b)$.

Figure 14.9 Realization of a Homogeneous Poisson Process

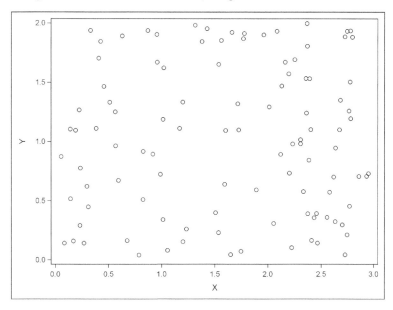

14.5   Simulating Data from an Inhomogeneous Poisson Process

For an inhomogeneous Poisson process, the intensity is a function, $\lambda(x, y)$. Again, assume that the domain is a rectangular region, A. Schabenberger and Gotway (2005) describe the following rejection algorithm (see Section 7.7) for simulating an inhomogeneous Poisson process, which they attribute to Lewis and Shedler (1979). The algorithm is known as a *thinning algorithm* because it first generates points from a dense homogeneous process and then accepts or rejects each point based on the intensity function, as follows:

1. Let $\lambda_0 = \max_A \lambda(x, y)$. Simulate points from a homogeneous Poisson process with parameter λ_0.

2. For each generated point s, keep the point with probability $\lambda(s)/\lambda_0$.

The second step is the thinning step. To determine whether to keep the point s_i, you can call the RAND function to sample from a Bernoulli distribution with probability $\lambda(s_i)/\lambda_0$. The following SAS/IML statements implement this thinning algorithm and simulate data from an inhomogeneous Poisson process. Figure 14.10 shows one realization of the inhomogeneous Poisson process with the specified intensity function. Notice that the points are more dense near the origin than near the point $(3, 2)$.

```
/* intensity of an inhomogeneous Poisson process */
start Intensity(x,y);
   return( 100/((x+y)##2 +1) );
finish;

/* simulate inhomogeneous Poisson process
   lambda0 is intensity of underlying homogeneous Poisson process */
start InhomogPoissonProcess(lambda0, a, b);
   u = HomogPoissonProcess(lambda0, a, b);
   lambda = Intensity(u[,1], u[,2]);
   r = rand("Bernoulli", lambda/lambda0);
   return( u[loc(r),] );
finish;

lambda0 = 100;  a = 3;  b = 2;                    /* max intensity */
z = InhomogPoissonProcess(lambda0, a, b);
```

Figure 14.10 Realization of an Inhomogeneous Poisson Process

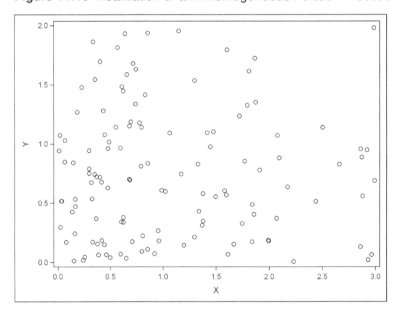

Exercise 14.7: Simulate a realization from an inhomogeneous Poisson process with $\lambda(x, y) = 100e^{-x}$. Visualize the resulting points.

14.6   Simulating Data from a Regular Process

A *regular process* (sometimes called a *hard-core process* (Illian et al. 2008)) generates points that have a more regular structure than those from a homogeneous Poisson process. The basic requirement is that no two points can be closer than some specified distance δ.

Schabenberger and Gotway (2005) describe several regular processes. One method is known as the Matérn model II. The model begins with points from a homogeneous Poisson process but thins the points to eliminate points that are closer than δ to some previously encountered point, as follows:

1. Simulate points from a homogeneous Poisson process.

2. Assign each point an "age," which is a uniform random value.

3. Specify a distance, δ. For each point, s, keep the point if the circle of radius δ centered at s does not contain any "older" points.

The idea of thinning due to age is motivated by the way that some trees grow. Trees that are older outcompete younger trees that are nearby. In a mature forest, the trees tend to be spaced apart by a certain distance.

You can use the SAS/IML language and the HOMOGPOISSONPROCESS function from Section 14.4 to simulate a regular point process. The algorithm requires knowing the distance between each pair of points. You can use the DISTANCE function in SAS/IML to compute the matrix of pairwise distances between a set of points. The DISTANCE function was introduced in SAS/IML 9.3. If you are running an earlier version, then use the EUCLIDEANDISTANCE function that is provided in Appendix A, "A SAS/IML Primer." The following statements implement the algorithm:

```
/* Simulate regular point process by thinning  */
proc iml;
call randseed(123);
a = 3; b = 2;
load module=(Uniform2D HomogPoissonProcess);   /* load from storage */

u = HomogPoissonProcess(100/6, a, b);
n = nrow(u);
d = distance(u);                /* matrix of pairwise distances     */
age = j(n,1);                   /* give points random "age" in [0,1] */
call randgen(age, "Uniform");
delta = 0.5;                    /* thin points closer than delta    */
d[do(1, n##2, n+1)] = .;        /* assign diagonal to missing       */

/* Sort points (and distances) according to age */
call sortndx(ndx, age, 1, 1);   /* index for age (desc order)   */
z = u[ndx,];                    /* contains the sorted points   */
d = d[ndx, ndx];

/* Retain s if no "older" point within distance delta */
keep = j(n,1,0);                         /* 0/1 indicator variable       */
keep[1] = 1;                             /* keep the oldest point        */
do i = 2 to n;
   v = d[i, 1:i-1];                      /* distances of older points    */
   j = xsect(loc(v < delta),            /* older points that are close  */
             loc(keep) );               /* points that are retained     */
   keep[i] = (ncol(j)=0);               /* retain this point?           */
end;
w = z[loc(keep), ];
```

Sorting the points according to their "age" makes the bookkeeping easier in the algorithm. Keep the oldest point. For each point, s, you only have to look at the distances between s and the points that are older than s. If an older point is within distance δ of s and that point has been kept at an earlier step, then do not keep s. Otherwise, keep s. At the end of the algorithm, the matrix **w** contains the points that were not thinned out.

Figure 14.11 shows a realization of a regular process with $\delta = 0.5$. The circular markers are the points that are retained by the process. The faint points that are marked by an X are points that were generated by the Poisson process, but were thinned (removed) by the algorithm.

Figure 14.11 Realization of a Regular Process

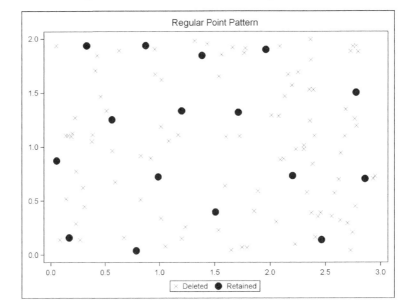

14.7   Other Techniques for Simulating Data from a Point Process

As shown in the previous section, thinning a point process to obtain a new process is a powerful technique for simulating spatial point processes. Schabenberger and Gotway (2005, p. 123) describe the following three kinds of thinning:

p**-thinning** is a method of uniformly thinning points. Given a probability, p, you retain each point with probability p and remove it with probability $1 - p$.

$p(s)$**-thinning** is a method of thinning points in a nonuniform way. Given a probability function, $p(s)$, you retain each point with probability given by its location. Notice that $p(s)$-thinning is the method used in Section 14.5 to simulate inhomogeneous Poisson point patterns.

π**-thinning** This thinning function is the realization of a spatial random field. Use the methods in Section 14.2 to obtain a realization of a field at the original points. Use that realization as the $p(s)$ function to carry out $p(s)$-thinning.

Schabenberger and Gotway (2005, p. 125) also describe how to generate a *clustered process*. Start with a Poisson process. The generated points are called the "parents." For each parent, generate "children" according to a bivariate distribution function centered at the parent. A clustered process might be used to model, for example, mature oak trees (the parents) that drop acorns around their base, which grow into new oak trees.

Exercise 14.8: Implement p-thinning. Generate points according to a homogeneous Poisson process, and generate a Bernoulli variable that determines whether each point is kept. Visualize the resulting point process when $p = 0.5$.

Exercise 14.9: Implement a simulation of a clustered process. Generate points according to a homogeneous Poisson process. For each point s_i, let n_i be the value of a Poisson random variable with mean 10 (see Section 2.4.6) and generate n_i new points according to a bivariate normal distribution with a diagonal covariance matrix $\mathrm{diag}(0.02, 0.01)$.

14.8   References

Illian, J., Penttinen, A., Stoyan, H., and Stoyan, D. (2008), *Statistical Analysis and Modelling of Spatial Point Patterns*, Hoboken, NJ: John Wiley & Sons.

Kolovos, A. (2010), "Everything in Its Place: Efficient Geostatistical Analysis with SAS/STAT Spatial Procedures," in *Proceedings of the SAS Global Forum 2010 Conference*, Cary, NC: SAS Institute Inc.
URL http://support.sas.com/resources/papers/proceedings10/337-2010.pdf

Lewis, P. A. W. and Shedler, G. S. (1979), "Simulation of Nonhomogeneous Poisson Processes by Thinning," *Naval Research Logistics Quarterly*, 26, 403–413.

Schabenberger, O. and Gotway, C. A. (2005), *Statistical Methods for Spatial Data Analysis*, Boca Raton, FL: Chapman & Hall/CRC.

Chapter 15
Resampling and Bootstrap Methods

Contents

15.1   An Introduction to the Bootstrap

Bootstrap methods are nonparametric techniques that simulate data directly from a sample. These techniques are often used when no statistical model of the data is evident. Introductory references for bootstrap techniques include Efron and Tibshirani (1993) and Davison and Hinkley (1997).

Bootstrap methods use the fact that the empirical CDF (ECDF) of the data is an approximation to the underlying distribution. Consequently, the basic bootstrap method is essentially an inverse CDF technique (see Section 7.4.2) for sampling from the ECDF. The basic bootstrap technique is also called the *simple bootstrap* or the *naive bootstrap*.

This chapter presents efficient techniques for implementing the simple bootstrap in SAS software. This chapter also presents two variations of the simple bootstrap. The so-called *parametric bootstrap* is actually a classical technique that predates the introduction of the bootstrap. In the parametric bootstrap, you fit a parametric distribution to the data and then sample from this parametric distribution. The second variation is called the *smooth bootstrap*. In the smooth bootstrap, you randomly select observations from the data and add random noise.

The purpose of this chapter is to demonstrate a few basic techniques for bootstrapping. This chapter is far from comprehensive. See Cassell (2007) and Cassell (2010) for additional discussion of bootstrapping in SAS.

15.2   Resampling Techniques in SAS Software

There are three general techniques in SAS software for resampling data:

The DATA Step: You can use the POINT= option in the SET statement to randomly select observations from a SAS data set. For many data sets you can use the SASFILE statement to read the entire sample data into memory, which improves the performance of random access.

The SURVEYSELECT Procedure: You can use the SURVEYSELECT procedure to randomly select observations according to several sampling schemes.

SAS/IML Software: You can use built-in or user-defined functions to resample from the data.

This chapter uses a subset of Fisher's famous iris data to illustrate the bootstrap technique. Execute the following statements to generate the example data:

```
data Virginica(drop=Species);
set Sashelp.Iris(where=(Species="Virginica"));
run;
```

The Virginica data contain measurements of 50 flowers of the species *Iris virginica*. The primary variable that is studied in this chapter is the SepalLength variable, which measures the length of the sepal (in millimeters) of each flower.

Suppose that you are interested in understanding the sampling variation for the skewness and kurtosis of the SepalLength variable. You can use PROC MEANS to compute point estimates, as shown in Figure 15.1.

```
proc means data=Virginica nolabels Skew Kurt;
   var SepalLength;
run;
```

Figure 15.1 Sample Skewness and Kurtosis

The MEANS Procedure

Analysis Variable : SepalLength	
Skewness	Kurtosis
0.1180151	0.0329044

PROC MEANS does not support computing the standard error for these statistics, nor does it support confidence intervals for the parameters. However, you can use bootstrap methods to estimate these quantities. Implementing a bootstrap method requires three main steps:

1. Compute the statistic of interest on the original data, as in Figure 15.1.

2. Resample B times from the data to form B bootstrap samples.

3. Compute the statistic on each resample.

The statistics that are computed from the resampled data form a *bootstrap distribution*. The bootstrap distribution is an estimate of the sampling distribution of a statistic. In particular, the standard deviation of the bootstrap distribution is an estimate for the standard error of the statistic. Percentiles of the bootstrap distribution are a simple way to estimate confidence intervals.

Bootstrap confidence intervals are especially useful when the alternative approach produces a symmetric confidence interval such as $\hat{s} \pm z_{\alpha/2}SE$, where $\hat{s}$ is a point estimate, $z_{\alpha/2}$ is a standard normal quantile, and SE is an estimate of the standard error. These symmetric confidence intervals are often based on asymptotic normality of the sampling distribution of the statistic. For small samples, the distribution might not be symmetric, much less normally distributed. In general, bootstrap estimates of a confidence interval are not symmetric about the point estimate.

Step 1 of the algorithm is not strictly necessary, but it is useful for studying bias in the estimate. It is also used for permutation tests and hypothesis tests: a p-value for the observed statistic is estimated by using a bootstrap distribution that is generated by resampling according to the null hypothesis.

You can use many of the tips and techniques in this book to implement bootstrap methods in SAS software. The main difference between simulation and bootstrapping is that simulation draws random samples from a model of the population, whereas bootstrap samples are created by sampling with replacement directly from the data.

15.3 Resampling with the DATA Step

The simplest way to sample with replacement in the DATA step is to use the POINT= option in the SET statement. The following DATA step implements a resampling algorithm:

```
%let MyData = Virginica;
%let NumSamples = 5000;

sasfile &MyData load;                                        /* 1     */
data BootDS(drop=i);
call streaminit(1);
do SampleID = 1 to &NumSamples;                             /* 2     */
   do i = 1 to NObs;                                        /* 4     */
      choice = ceil(NObs * rand("Uniform"));                /* 5     */
      set &MyData point=choice nobs=NObs;                   /* 3, 6 */
      output;
   end;
end;
STOP;                                                       /* 7     */
run;
sasfile &MyData close;                        /* release memory */
```

The DATA step uses the following techniques, which are indicated by comments:

1. Provided that the data set is not too large, use the SASFILE statement to load the data into memory, which speeds up random access.

2. The SampleID variable is used to identify each sample.

3. SAS sets the value of the NObs variable to be the number of observations. This value is set before the DATA step runs. Consequently, the value of NObs is available throughout the DATA step, and in particular on statements that execute prior to the SET statement.

4. Randomly sample NObs observations.

5. Select a random integer in the range 1–NObs, as discussed in Section 2.4.4.

6. Use the POINT= option to read the selected observation.

7. Stop processing when all resamples are output.

The BootDS data set contains 5,000 samples, each with 50 observations. The samples are identified by the SampleID variable. This technique, in which an entire row is output at each iteration, is known as *case resampling*.

If you are interested only in one or two variables, then you can use a KEEP statement to output only certain variables. Alternatively, you can use an ARRAY statement to cache the observations for certain variables in an array. Random access of elements in an array is usually more efficient than random access of observations in a data set. To use this technique, you have to know how many observations are in the data set. The following DATA step stores that information in a macro variable:

```
data _null_;
   call symput('N', NObs);
   if 0 then set &MyData nobs=NObs;
   STOP;
run;
```

The following DATA step resamples from the SepalLength variable 5,000 times. The main advantage over the previous method is that this technique reads each observation only once.

```
%let VarName = SepalLength;
data BootArray(keep= SampleID &VarName);
array arr_y[&N] _temporary_;

do i = 1 to NObs;                          /* read data one time   */
   set &MyData point=i nobs=NObs;          /* store obs in array    */
   arr_y[i] = &VarName;
end;

do SampleID = 1 to &NumSamples;            /* resampling algorithm */
   do i = 1 to NObs;
      choice=ceil(NObs * rand("Uniform"));
      &VarName = arr_y[choice];            /* get value from array */
      output;
   end;
end;
STOP;
run;
```

The program uses an array to hold an entire variable in memory. This technique is easy to implement for one variable. The disadvantage of this method is that it is memory intensive and that you need an array for every variable that you want to bootstrap. (Alternatively, you could use a double array.)

You can use BY-group processing to analyze the BootDS or the BootArray data in the usual way. For example, the following statement computes the bootstrap estimates of skewness and kurtosis for the SepalLength variable:

```
/* compute bootstrap estimate on each bootstrap sample */
proc means data=BootDS noprint;
   by SampleID;
   var SepalLength;
   output out=OutStats skew=Skewness kurt=Kurtosis;
run;
```

The OutStats data set contains 5,000 observations. Each observation is the sample skewness and kurtosis of one of the bootstrap resamples. The union of these statistics is the bootstrap distribution of the statistic. The following statements construct a scatter plot that shows the joint distribution of the sample skewness and kurtosis on the bootstrap resamples:

```
proc sgplot data=OutStats;
   title "Bootstrap Estimates of Skewness and Kurtosis";
   scatter x=Skewness y=Kurtosis / transparency=0.7;
   refline 0.118 / axis=x;       /* reference line at observed values */
   refline 0.033 / axis=y;
   label Skewness= Kurtosis=;
run;
```

Figure 15.2 Bootstrap Distribution of the Sample Skewness and Kurtosis

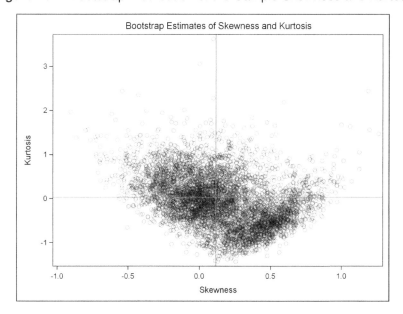

You can also request summary statistics for the bootstrap distribution. The mean is the bootstrap estimate of the parameter. The standard deviation is the bootstrap estimate for the standard error. The $\alpha/2$ and $1 - \alpha/2$ quantiles of the bootstrap distribution form an approximate $100(1 - \alpha)\%$ confidence interval. The P5 and P95 options in the PROC MEANS statement make it easy to compute a 90% confidence interval, as shown in Figure 15.3.

```
proc means data=OutStats nolabels N Mean StdDev P5 P95;
   var Skewness Kurtosis;
run;
```

Figure 15.3 Descriptive Statistics for Skewness and Kurtosis

The MEANS Procedure

Variable	N	Mean	Std Dev	5th Pctl	95th Pctl
Skewness	5000	0.1534985	0.3206183	-0.3482734	0.6766532
Kurtosis	5000	-0.0378410	0.5900981	-0.9417224	0.9857613

To compute a 95% confidence interval, use PROC UNIVARIATE to compute the 2.5 and 97.5 percentiles of the data, as shown in the following statements. The result is shown in Figure 15.4.

```
proc univariate data=OutStats noprint;
   var Skewness Kurtosis;
   output out=Pctl95 pctlpts =2.5  97.5   pctlname=P025 P975
               pctlpre =Skew_ Kurt_ mean=SkewMean KurtMean;
run;

proc print data=Pctl95 noobs; run;
```

Figure 15.4 Means and 95% Confidence Intervals

SkewMean	KurtMean	Skew_P025	Kurt_P025	Skew_P975	Kurt_P975
0.15350	-0.037841	-0.44484	-1.05536	0.77265	1.20269

Exercise 15.1: Draw histograms of the bootstrap distributions of the skewness and kurtosis.

Exercise 15.2: The following statements define a macro variable that records the elapsed time:

```
%let t0 = %sysfunc(datetime());
   /* put computation here */
%let t1 = %sysfunc(datetime());
%let elapsedTime = %sysevalf(&t1-&t0);
```

Compare the times for each DATA step technique to generate 5,000 resamples of the Virginica data.

15.4   Resampling with the SURVEYSELECT Procedure

The previous section shows that for the simplest sampling scheme (sampling with replacement), it is easy to use the DATA step to program a resampling scheme. However, experienced SAS programmers use the SURVEYSELECT procedure for resampling from data for the following reasons:

- It is about as fast as the DATA step array method.

- It supports many sampling schemes and can be used for *jackknife methods*.

- It requires no programming.

- It manufactures a frequency variable, which can decrease the size of the output data set.

For moderate size data, use the SASFILE statement (see Section 15.3) to load the data into RAM. This technique is not shown for the example in this section.

15.4.1   Resampling without Creating a Frequency Variable

The following statements, which are described later, use PROC SURVEYSELECT to generate 5,000 resamples of observations in the Virginica data set, which was created in Section 15.2:

```
%let MyData = Virginica;
%let NumSamples = 5000;
proc surveyselect data=&MyData NOPRINT seed=1       /* 1 */
     out=BootSS(rename=(Replicate=SampleID))        /* 2 */
     method=urs samprate=1                          /* 3 */
     reps=&NumSamples                               /* 4 */
     outhits;                                       /* 5 */
run;
```

The call to PROC SURVEYSELECT does the following:

1. The SEED= option specifies the seed value for random number generation. If you specify a zero seed, then omit the NOPRINT option so that the value of the chosen seed appears in the procedure output.

2. The BY-group variable in the output data set is named Replicate. For consistency, rename this variable to SampleID.

3. The METHOD=URS option specifies unrestricted random sampling, which means sampling with replacement and with equal probability. The SAMPRATE=1 option specifies that each resample is the same size as the original data.

4. The REPS= option specifies how many bootstrap resamples you want to generate.

5. The OUTHITS options specifies that the output data set contains $N \times$ NumSamples observations, where N is the number of observations in the original data set (assuming that SAMPRATE=1). If a record is selected three times, then the output data set contains three (consecutive) copies of that record. In contrast, if you omit the OUTHITS option, then the output data set has fewer observations and the NumberHits variable contains the number of times that each record was selected. Consequently, you can refer to the NumberHits variable in the FREQ (or WEIGHT) statement of procedures that support the FREQ statement.

The following statements create bootstrap distributions for the skewness and kurtosis. The MEANS procedure produces Figure 15.5, which shows the mean, standard deviation, and middle 90% interval for the bootstrap distribution for the skewness and kurtosis. Compare these values with Figure 15.3.

```
proc means data=BootSS noprint;
   by SampleID;
   var SepalLength;
   output out=OutStats skew=Skewness kurt=Kurtosis;
run;

proc means data=OutStats nolabels N Mean StdDev P5 P95;
   var Skewness Kurtosis;
run;
```

Figure 15.5 Descriptive Statistics for Bootstrap Distributions

The MEANS Procedure

Variable	N	Mean	Std Dev	5th Pctl	95th Pctl
Skewness	5000	0.1534985	0.3206183	-0.3482734	0.6766532
Kurtosis	5000	-0.0378410	0.5900981	-0.9417224	0.9857613

15.4.2   Resampling and Creating a Frequency Variable

You can reduce the size of the output data set by omitting the OUTHITS option in the SURVEYSE-LECT procedure. If you omit the option, then the NumberHits variable in the output data set contains the frequency for each observation. Consequently, you can analyze the output data as follows:

```
proc surveyselect data=&MyData NOPRINT seed=1
      out=BootSSFreq(rename=(Replicate=SampleID))
      method=urs samprate=1 reps=&NumSamples;
run;

proc means data=BootSSFreq noprint;
   by SampleID;
   freq NumberHits;
   var SepalLength;
   output out=OutStats2 skew=Skewness kurt=Kurtosis;
run;
```

Exercise 15.3: Run the preceding statements. Analyze the bootstrap distributions and compare your results with Figure 15.5. Do you obtain the same results if you use the same SEED= option?

15.5   Resampling Univariate Data with SAS/IML Software

SAS/IML 12.1 supports the SAMPLE function (see Section 2.6.2) that you can use to sample (with or without replacement) from a finite set. Prior to SAS/IML 12.1, you can use the SAMPLEREPLACE module, which is included in Appendix A. The following SAS/IML program uses the SKEWNESS and KURTOSIS modules, which are defined in Appendix A, to compute the bootstrap distributions for the skewness and kurtosis of the SepalLength variable. The bootstrap estimates are shown in Figure 15.6. They are similar to the estimates given in Figure 15.5. The SAS/IML QNTL function can compute any quantile, so you can use the parameters {0.025 0.975} to obtain a 95% confidence interval.

```
%let MyData = Virginica;
%let NumSamples = 5000;

/* Basic bootstrap to explore variation of skewness and kurtosis */
proc iml;
call randseed(12345);
load module=(Skewness Kurtosis); /* load SampleReplace if necessary */
use &MyData;
read all var {SepalLength} into x;
close &MyData;

/* get all bootstrap resamples with a single call */
/*   s = SampleReplace(x, nrow(x), &NumSamples); */ /* prior to 12.1 */
s = sample(x, &NumSamples // nrow(x));        /* 50 x NumSamples     */
M = Skewness(s) // Kurtosis(s);               /* bootstrap statistics */
M = M`;                                        /* NumSamples x 2      */

means = mean(M);                      /* summarize bootstrap distribution */
call qntl(q, M, {0.05 0.95});
s = means` || q`;
VarNames = {"Skewness" "Kurtosis"};
StatNames = {"Mean" "P5" "P95"};
print s[format = 9.5 r=VarNames c=StatNames];
```

Figure 15.6 Means and Confidence Intervals, Computed in PROC IML

s			
	Mean	**P5**	**P95**
Skewness	0.16774	-0.34522	0.70022
Kurtosis	-0.04636	-0.95937	0.98463

15.6   Resampling Multivariate Data with SAS/IML Software

The previous section constructs bootstrap samples of a single variable and uses the samples to construct the bootstrap distribution of the skewness and kurtosis statistics. This is done without writing a single loop. The program uses a SAS/IML matrix to hold all of the bootstrap samples, one in each column. The SAS/IML language enables you to compute the sample skewness and kurtosis statistics for each column in a vectorized manner.

As discussed in Section 4.5.3, if you are generating multivariate samples, then it is often convenient to write a loop. Inside the loop you generate each multivariate sample and compute the corresponding statistic.

To illustrate this technique, consider the task of bootstrapping the correlation coefficients for the SepalLength, SepalWidth, and PetalLength variables in the Virginica data set. The goal is to estimate the distribution of the correlation coefficients, including a 95% confidence interval. As a check, you can use PROC CORR and Fisher's z transformation to generate the sample correlation and the 95% confidence intervals, which are shown in Figure 15.7:

```
/* compute sample correlations and Fisher 95% CI */
proc corr data=&MyData noprob fisher(biasadj=no);
   var SepalLength SepalWidth PetalLength;
   ods select FisherPearsonCorr;
run;
```

Figure 15.7 Sample Correlation Coefficients and 95% Confidence Intervals

The CORR Procedure

Pearson Correlation Statistics (Fisher's z Transformation)							
Variable	With Variable	N	Sample Correlation	Fisher's z	95% Confidence Limits		p Value for H0:Rho=0
SepalLength	SepalWidth	50	0.45723	0.49380	0.204966	0.652529	0.0007
SepalLength	PetalLength	50	0.86422	1.30980	0.771454	0.921017	<.0001
SepalWidth	PetalLength	50	0.40104	0.42489	0.138115	0.611168	0.0036

You can use bootstrap methods to approximate the sampling distribution for the correlation coefficients. Each bootstrap sample is a 50×3 matrix that is generated by resampling with replacement from the 50 rows in the data. For each sample, you can compute the Pearson correlation coefficients for the three variables. The following SAS/IML program stores the correlation coefficients in rows of the **rho** matrix. Figure 15.8 displays some descriptive statistics for the bootstrap distributions.

```
/* bootstrap of MV samples */
proc iml;
call randseed(12345);
use &MyData;
read all var {"SepalLength" "SepalWidth" "PetalLength"} into X;
close &MyData;

/* Resample from the rows of X. Generate the indices for
   all bootstrap resamples with a single call */
N = nrow(X);
/* ndx = SampleReplace(1:N, &NumSamples, N); */
ndx = Sample(1:N, N // &NumSamples);        /* NumSamples x N       */

rho = j(&NumSamples, ncol(X));      /* allocate for results        */
do i = 1 to &NumSamples;
   rows = ndx[i, ];                 /* selected rows for i_th sample */
   Y = X[rows, ];                   /* the i_th sample              */
   c = corr(Y);                     /* correlation matrix           */
   rho[i, ] = c[{2 3 6}]`;          /* upper triangular elements    */
end;
```

```
means = mean(rho);                    /* summarize bootstrap distrib   */
call qntl(q, rho, {0.025 0.975});
s = means` || q`;
varNames = {"p12" "p13" "p23"};
StatNames = {"Mean" "P025" "P975"};
print s[format = 9.5 r=VarNames c=StatNames];
```

Figure 15.8 Bootstrap Estimates of Correlations and 95% Confidence Intervals

	s		
	Mean	P025	P975
p12	0.44945	0.17070	0.67619
p13	0.86168	0.77485	0.91961
p23	0.39699	0.10017	0.66514

The means of the bootstrap distributions are close to the sample correlations, which are shown in Figure 15.7. The confidence intervals are also close to the Fisher confidence limits. The main difference is the wider confidence interval for p23, which is the correlation between SepalWidth and PetalLength.

Exercise 15.4: Write the `rho` matrix to a SAS data set and use the MATRIX statement in the SGSCATTER procedure to visualize the joint distribution of the correlation coefficients. Use the DIAGONAL= option to add histograms and kernel density estimates for each correlation coefficient.

15.7   The Parametric Bootstrap Method

What some people call the *parametric bootstrap* is nothing more than the process of fitting a model distribution to the data and simulating data from the fitted model.

The first step is to choose a model for the data. If you have domain-specific knowledge of the process that generates the data, then that knowledge might suggest a model. For example, exponential models and Poisson models are often chosen because you assume that some process occurs at a constant rate.

If you do not have domain-specific knowledge, then choosing a distribution that models the data can be difficult. Chapter 16 describes how to use the moment-ratio diagram to choose candidate distributions from common "named" distributions. Alternatively, the same chapter describes how to use moment-matching to construct a distribution that matches the first four central moments of the sample data.

After you choose a model, you can use various SAS procedures to fit the model parameters. For continuous univariate data, PROC UNIVARIATE supports fitting a variety of common distributions. See Section 16.6 for an example. For more sophisticated models, there are many SAS procedures that fit parameters to data including PROC MODEL in SAS/ETS software and PROC NLIN in SAS/STAT software.

The last step is simulation from the model, which is discussed throughout this book.

15.8   The Smooth Bootstrap Method

In Section 15.4, the BootSS data set was created, which contains 5,000 resamples from the Virginica data. The following statements plot the bootstrap distribution for the median of the SepalLength variable. The distribution is shown in Figure 15.9.

```
proc means data=BootSS noprint;
   by SampleID;
   var SepalLength;
   output out=OutMed median=Median;
run;

proc univariate data=OutMed;
   histogram Median / kernel;
   ods select histogram;
run;
```

Figure 15.9 Bootstrap Distribution of the Sample Median

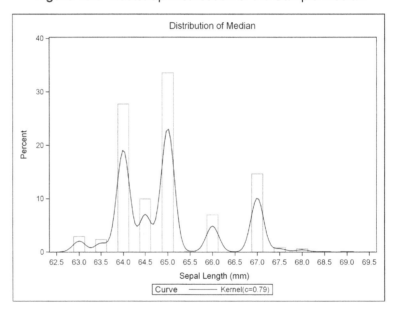

Figure 15.9 shows that the bootstrap distribution for the median is not continuous. However, the *true* sampling distribution of the median *is* continuous, so what is going on?

The problem is that the median (or any quantile) of a bootstrap resample can only attain a small number of values. For a sample of size N, it is a combinatorial fact that the bootstrap median can only assume at most $2N - 1$ possible values: the N sample values and the $N - 1$ midpoints between adjacent ordered values. In practice, the bootstrap median attains a small number of possible values determined by data values near the sample median, as shown in Figure 15.9.

One way to overcome this problem is to add a small amount of random noise to each data point that is selected during the bootstrap resampling. This is equivalent to sampling from a kernel density estimate rather than sampling from the ECDF.

The question arises, if you intend to add a random quantity of the form $\epsilon_i \sim N(0, \lambda)$, where λ is the standard deviation parameter, then what value should you choose for λ? Some researchers suggest choosing λ to be an estimate of the standard error of the statistic. Others might prefer to use the bandwidth of the kernel density estimate of the original data.

A bootstrap estimate of the standard error of the median is about 1.11. Because the UNIVARIATE procedure reports a standardized bandwidth, the following statements use PROC KDE to compute the bandwidth. See Figure 15.10.

```
%let MyData = Virginica;
%let VarName = SepalLength;
proc kde data=&MyData;
   univar SepalLength / method=SJPI unistats;
   ods select UnivariateStatistics;
run;
```

Figure 15.10 Summary Statistics and Bandwidth for Sample Data

The KDE Procedure

Univariate Statistics	
	SepalLength
Mean	65.88
Variance	40.43
Standard Deviation	6.36
Range	30.00
Interquartile Range	7.00
Bandwidth	2.59

Regardless of how you obtain a value for λ, the smooth bootstrap proceeds as follows. If the data value x_i is selected for inclusion into a bootstrap sample, then use the value $x_i + \epsilon_i$, where $\epsilon_i \sim N(0, \lambda)$. The following SAS/IML program implements the smooth bootstrap method:

```
proc iml;
/* Smooth bootstrap.
   Input: A is an input vector with N elements.
   Output: (B x N) matrix. Each row is a sample.
   Prior to SAS/IML 12.1, use the SampleReplace module */
start SmoothBootstrap(x, B, Bandwidth);
   N = nrow(x) * ncol(x);
   /* s = SampleReplace(x, B, N); */        /* prior to SAS/IML 12.1 */
   s = Sample(x, N // B);                    /* B x N matrix     */
   eps = j(B, N);                            /* allocate vector */
   call randgen(eps, "Normal", 0, Bandwidth);  /* fill vector      */
   return( s + eps );                        /* add random term */
finish;
```

You can call the SMOOTHBOOTSTRAP module on the SepalLength variable in the Virginica data. The SMOOTHBOOTSTRAP module returns each bootstrap sample in a row, so transpose the return value before computing the median. The following program uses the Sheather-Jones bandwidth estimate ($\lambda = 2.59$) that is shown in Figure 15.10 to produce the histogram that is shown in Figure 15.11. Compare Figure 15.11 with Figure 15.9. The smooth bootstrap is a much better estimate of the sampling distribution of the median.

```
use &MyData;  read all var {SepalLength} into x;  close &MyData;

call randseed(12345);
y = SmoothBootstrap(x, &NumSamples, 2.59);        /* SJPI bandwidth */
Median = Median(y`);                       /* smooth bootstrap estimates */
create Smooth var {"Median"}; append; close Smooth;
quit;

proc univariate data=Smooth;
   histogram Median / kernel;
   ods select histogram;
run;
```

Figure 15.11 Smooth Bootstrap Distribution of the Sample Median

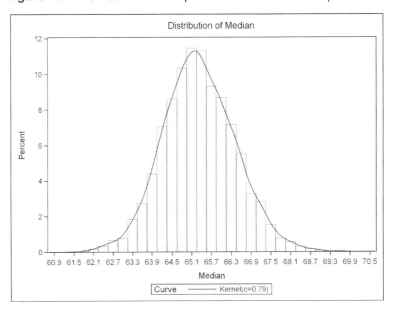

Exercise 15.5: Implement the smooth bootstrap in the DATA step by adding a small random normal variate to each observation in the OutMed data set, which is created in Section 15.4.

15.9   Computing the Bootstrap Standard Error

The bootstrap standard error is estimated by computing the standard deviation of the bootstrap distribution. The usual sample estimate of variance has a divisor of $N - 1$ for a sample of size N. However, the bootstrap estimate of variance requires using N as a divisor (Davison and Hinkley

1997). This means that to compute a bootstrap estimate of the variance of a statistic (or any related statistic, such as a Student's t confidence interval), then you need to estimate the variance by using a slightly different formula.

Both PROC MEANS and PROC UNIVARIATE support a VARDEF= option in the PROC statement. The default value is VARDEF=DF, which results in a variance divisor of $N - 1$. You can correct the variance estimate by multiplying it by the factor $(N - 1)/N$. Alternately, you can use the VARDEF=N option.

15.10   Computing Bootstrap Confidence Intervals

In this chapter, simple quantiles of the bootstrap distribution are used to construct confidence intervals. (For a significance level α, a basic confidence interval is formed from the $\alpha/2$ and $1 - \alpha/2$ quantiles.) This simple method performs well for quantiles and for unbiased statistics.

However, for biased statistics, the simple method amplifies the bias. There are popular techniques that can be used to correct for bias (Efron and Tibshirani 1993). The bias-corrected (BC) confidence intervals adjust the quantiles. That is, instead of using the $\alpha/2$ and $1 - \alpha/2$ quantiles, new quantiles are computed that depend on the estimate of the bias. Another technique is the bias-corrected and accelerated (BC_a) confidence interval. A good overview and discussion of these techniques are available in the article, comments, and rejoinder by DiCiccio and Efron (1996).

In SAS, several bootstrap techniques are provided by a series of macros that are provided by SAS. The %BOOT and %BOOTCI macros provide bootstrap methods and several kinds of confidence intervals. These macros and documentation are available at `support.sas.com/kb/24/982.html`.

15.11   References

Cassell, D. L. (2007), "Don't Be Loopy: Re-sampling and Simulation the SAS Way," in *Proceedings of the SAS Global Forum 2007 Conference*, Cary, NC: SAS Institute Inc.
URL `http://www2.sas.com/proceedings/forum2007/183-2007.pdf`

Cassell, D. L. (2010), "BootstrapMania!: Re-sampling the SAS Way," in *Proceedings of the SAS Global Forum 2010 Conference*, Cary, NC: SAS Institute Inc.
URL `http://support.sas.com/resources/papers/proceedings10/268-2010.pdf`

Davison, A. C. and Hinkley, D. V. (1997), *Bootstrap Methods and Their Application*, Cambridge: Cambridge University Press.

DiCiccio, T. J. and Efron, B. (1996), "Bootstrap Confidence Intervals," *Statistical Science*, 11, 189–212.

Efron, B. and Tibshirani, R. J. (1993), *An Introduction to the Bootstrap*, New York: Chapman & Hall.

Chapter 16
Moment Matching and the Moment-Ratio Diagram

Contents

16.1 Overview of Simulating Data with Given Moments

Given data, how can you simulate additional samples that have the same distributional properties? There are several techniques:

- You can sample from the empirical distribution of the data, which is precisely the basic bootstrap algorithm, as discussed in Chapter 15, "Resampling and Bootstrap Methods."

- You can use a well-known "named" parametric distribution to model the data. You can then simulate the data by drawing random samples from the fitted distribution.

- You can model the data as a finite mixture of distributions. The FMM procedure in SAS/STAT software can fit a wide range of mixture models to data. After you have fit a model, you can use the ideas in Section 7.5 to simulate data from the model.

- For univariate data, you can choose a flexible system of distributions such as the Pearson system (Ord 2005), the Burr system (Burr 1942; Rodriguez 2005), or the Johnson system (Johnson 1949; Bowman and Shenton 1983; Slifker and Shapiro 1980).

- You can use Fleishman's method (Fleishman 1978) and its multivariate generalization (Vale and Maurelli 1983) to construct a distribution whose population mean, variance, skewness, and kurtosis match the corresponding sample statistics for the data.

These approaches are summarized in Figure 16.1. This chapter describes how to use the Johnson system and the Fleishman transformation; the other methods have been discussed previously. Tadikamalla (1980) gives a concise summary of the advantages and disadvantages of the Johnson system and the Fleishman transformation.

Figure 16.1 Some Methods to Simulate from Observed Data

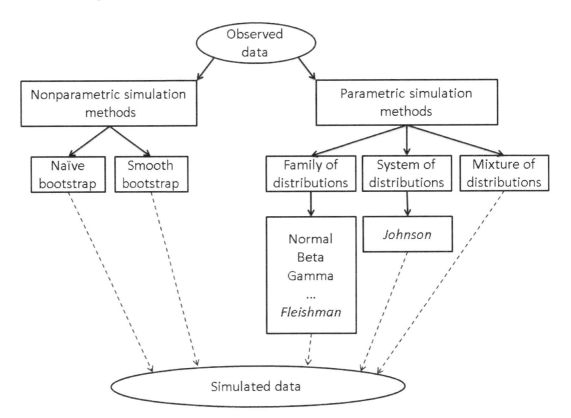

The idea of simulating data from a parametric distribution whose mean, variance, skewness, and kurtosis match the corresponding sample statistics for the data is called *moment matching.* This chapter describes how to use moment matching to simulate data from a distribution with a specified mean, variance, skewness, and kurtosis.

This chapter also describes a useful graphical tool, called the *moment-ratio diagram.* This diagram can help you to select candidate distributions to model the data and to compare simulations from different distributions. Traditionally the moment-ratio diagram is a theoretical tool for understanding the relationships of families and systems of distributions. This chapter shows how to use the moment-ratio diagram as a tool to organize simulation studies.

16.2   Moments and Moment Ratios

The *moments* of a distribution measure aspects of the distribution's location, scale, and shape. For a random variable X, let μ be the mean, which is the first (raw) moment. The *central moment of order* r is $\mu_r = E((X - \mu)^r)$, where E denotes the expected value operator. For example, the second central moment is the variance of the distribution, $\mu_2 = \sigma^2$.

The skewness and kurtosis are the third and fourth *standardized* central moments, respectively. The word "standardized" indicates that the deviation about the mean is divided by the standard deviation. The skewness is defined as

$$\gamma = E\left[\left(\frac{X - \mu}{\sigma}\right)^3\right] = \frac{\mu_3}{\sigma^3} = \frac{\mu_3}{\mu_2^{3/2}}$$

and the kurtosis is defined as

$$\kappa = E\left[\left(\frac{X - \mu}{\sigma}\right)^4\right] = \frac{\mu_4}{\sigma^4} = \frac{\mu_4}{\mu_2^2}$$

Because the skewness and kurtosis are ratios of centralized moments, they are referred to as *moment ratios*. The kurtosis of the normal distribution is 3, so many researchers use the quantity $\kappa - 3$, which is known as the *excess kurtosis* or the *coefficient of excess*. In fact, many references drop the "excess" modifier altogether. For example, the statistic labeled "kurtosis" in the output of SAS procedures is actually an estimate of the excess kurtosis. When there is no chance for confusion, this book uses "kurtosis" to mean excess kurtosis. When clarity is needed, the modifiers "full" or "excess" are used.

In this chapter, the adjectives "central" and "standardized" are usually dropped and the phrase "the first four moments" is sometimes used to refer to the mean, variance, skewness, and kurtosis of a distribution.

Of course, there are sample statistics that estimate these quantities. The formulas that SAS uses are the same as in Kendall and Stuart (1977) and are given in the appendix "SAS Elementary Statistics Procedures" in the *Base SAS Procedures Guide*. A SAS/IML program that computes estimates is included in this book in Appendix A, "A SAS/IML Primer."

16.3   The Moment-Ratio Diagram

The moment-ratio diagram (Johnson, Kotz, and Balakrishnan 1994) shows the relationship between the skewness and the kurtosis for any univariate family of distributions for which the first four moments exist.

Many families include parameters (called *shape parameters*) that change the skewness (γ) and full kurtosis (κ) of the distribution. By plotting the locus of (γ, κ) values as the shape parameters vary over their possible values, you can visualize the relationship between the skewness and kurtosis for each distribution. (In some references, the locus of (γ^2, κ) values is plotted instead.)

By convention, the moment-ratio diagram is shown "upside down," with the kurtosis axis pointing down. Also by convention, the full kurtosis, κ, is shown rather than the excess kurtosis, $\kappa - 3$. For convenience, the diagrams in this book include axes for both the full kurtosis and the excess kurtosis.

The theoretical skewness and kurtosis of a distribution cannot take on arbitrary values. The kurtosis and skewness must obey the relationship $\kappa \geq 1 + \gamma^2$. This defines the *feasible region*.

Table 16.1 shows the skewness and excess kurtosis for a small set of continuous distribution. The moment-ratios for these distributions are shown in Figure 16.2.

Table 16.1   Skewness and Excess Kurtosis for Some Continuous Distributions

Distribution	**Skewness (γ)**	**Excess Kurtosis ($\kappa - 3$)**
Beta(α, β)	$\dfrac{2(\beta-\alpha)\sqrt{\alpha+\beta+1}}{(\alpha+\beta+2)\sqrt{\alpha\beta}}$	$\dfrac{6[(\alpha-\beta)^2(\alpha+\beta+1)-\alpha\beta(\alpha+\beta+2)]}{\alpha\beta(\alpha+\beta+2)(\alpha+\beta+3)}$
Exponential	2	6
Gamma(α)	$2/\sqrt{\alpha}$	$6/\alpha$
Gumbel	1.14	2.4
Lognormal(μ, σ)	$(e^{\sigma^2}+2)\sqrt{e^{\sigma^2}-1}$	$e^{4\sigma^2}+2e^{3\sigma^2}+3e^{2\sigma^2}-6$
Normal	0	0
t_ν ($\nu > 4$)	0	$6/(\nu-4)$

Figure 16.2   Moment-Ratio Diagram for Some Continuous Distributions

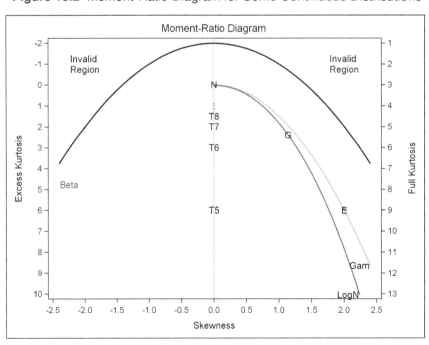

Figure 16.2 graphically demonstrates the following facts about the distributions in Table 16.1:

- The normal, Gumbel, and exponential distributions have no shape parameters. They are represented in the moment-ratio diagram as *points* labeled N, G, and E, respectively.

- The t distribution has a discrete shape parameter: the degrees of freedom. The t family is represented as a sequence of points that converges to the normal distribution. Like all symmetric distributions, these points lie along the $\gamma = 0$ axis.

- The lognormal and gamma distributions have one shape parameter and are represented as *curves*. For example, the locus of points for the gamma distribution (the upper curve at the edge of the beta region) is the parametric curve $(2/\sqrt{\alpha}, 6/\alpha)$ for $\alpha > 0$.

- The beta distribution has two shape parameters. It is represented by a *region* in the moment-ratio diagram.

The moment-ratio diagram is drawn by using the annotation facility of the SGPLOT procedure, which was introduced in SAS 9.3. You can use the annotation facility to draw arbitrary curves, regions, and text on a graph that is created by PROC SGPLOT. This book's Web site contains SAS programs that display the moment-ratio diagram for continuous distributions.

Figure 16.2 contains two vertical axes. The axis on the left indicates the excess kurtosis, whereas the axis on the right indicates the full kurtosis, κ. By convention, the axes point down.

Notice that the moment-ratio diagram does not fully describe a distribution because only the skewness and kurtosis are represented. For example, the fact that the Gumbel distribution lies on the lognormal curve does not imply that the Gumbel distribution is a member of the lognormal family, only that the distributions have the same skewness and kurtosis for some lognormal parameter value.

Figure 16.2 is a simple moment-ratio diagram. Vargo, Pasupathy, and Leemis (2010) present a more comprehensive moment-ratio diagram that includes 37 different distributions. The applications in this chapter do not require this complexity, but the diagram is reproduced in Figure 16.3 so that you can appreciate its beauty. The figure is reprinted with permission from the *Journal of Quality Technology©2010* American Society for Quality. No further distribution allowed without permission.

Exercise 16.1: Show that the Bernoulli distribution with parameter p exactly satisfies the equation $\kappa = 1 + \gamma^2$. In this sense, the Bernoulli distribution has the property that its kurtosis is as small as possible given its skewness.

Exercise 16.2: The moment ratios for the Gamma(4) distribution are close to those of the Gumbel family. Plot the density of the Gamma(4) distribution and compare its shape to that of several Gumbel densities. The PDF of the Gumbel(μ, σ) distribution is $f(x) = \exp(-z - \exp(-z))/\sigma$, where $z = (x - \mu)/\sigma$.

Figure 16.3 Comprehensive Moment-Ratio Diagram

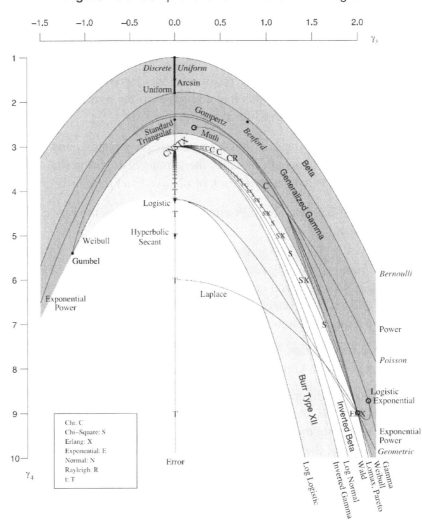

16.4   Moment Matching As a Modeling Tool

The problem of finding a flexible family (or system) of distributions that can fit a wide variety of distributional shapes is as old as the study of statistics. Between 1895 and 1916, Karl Pearson constructed a system of seven families of distributions that model data. Given a valid combination of the first four sample moments, the Pearson system provides a family that matches these four moments. Although the Pearson system is valuable for theory, practitioners prefer systems with fewer families such as the Johnson system.

A drawback of these systems is that they cannot be used to fit all distributions. They are primarily useful for fitting unimodal distributions, although the Johnson system and other parametric models can also be used to fit some bimodal distributions.

Although some modelers now favor nonparametric techniques, parametric families of distributions offer certain advantages such as the ability to fit a distributional form that involves a small number of parameters, some of which might have a practical interpretation. Consequently, the practice of fitting parametric distributions to data is still popular.

Moment matching is an attempt to match the shape of data by using four parameters that are easy to compute and relatively easy to interpret. The mean and variance estimate the expected value and the spread of the data, respectively. The skewness describes the asymmetry of the data. Positive skewness indicates that data are more prevalent in the right tail than in the left tail. The kurtosis is less intuitive, but for unimodal distributions it is generally interpreted as describing whether the distribution has a sharp peak and long tails (high kurtosis) or a low wide peak and short tails (low kurtosis).

Because the first two moments do not affect the shape of the data, the rest of this chapter is primarily concerned with finding distributions with a given skewness and kurtosis. If you use moment matching for small samples, keep in mind that "the estimates of these moments are highly biased for small samples" (Slifker and Shapiro 1980), and that they are very sensitive to outliers.

16.5 Plotting Variation of Skewness and Kurtosis on a Moment-Ratio Diagram

Now that the moment-ratio diagram has been explained, how can you apply it to the problem of simulating data? One application is to use the diagram to "identify likely candidate distributions" for data (Vargo, Pasupathy, and Leemis 2010, p. 6).

The simplest way to use the moment-ratio diagram is to locate the sample skewness and kurtosis of observed data on the diagram, and then look for "nearby" theoretical distributions. However, it is often not clear which distributions are "nearby" and which distributions are "far away." Nearness depends on the standard errors of the sample skewness and kurtosis. You can use bootstrap techniques to estimate the standard errors as shown in Section 15.3.

16.5.1 Example Data

The following DATA step creates 200 positive "observed" values that are used in this and subsequent sections. (Full disclosure: These data are fake.) Figure 16.4 shows summary statistics and Figure 16.5 shows a histogram and kernel density estimate of the data.

```
data MRData;
input x @@;
datalines;
 4.54 4.57 7.18 5.03 3.70   4.11 2.79 2.30 1.75 2.08 1.70 4.83 4.57 11.51
 2.36 6.47 3.86 7.14 6.96   3.59 5.81 5.66 7.07 2.29 4.42 1.01 6.49  2.59
 5.36 3.90 6.50 4.97 5.29   4.83 4.62 3.04 3.67 3.68 4.09 4.95 1.66  4.07
 4.31 2.20 2.29 6.38 3.58   4.11 2.50 2.94 1.47 6.77 9.54 2.14 2.84  3.25
 2.65 5.62 4.41 1.18 3.76   0.95 4.67 5.17 1.08 4.09 2.84 1.96 6.23  3.48
 5.41 6.17 7.71 2.84 2.32   4.40 3.21 2.22 0.56 3.53 3.03 1.35 1.97  1.61
 3.02 2.49 4.06 2.82 6.22  13.18 4.04 3.56 3.65 2.48 3.90 3.44 5.11  3.93
 1.69 6.12 2.75 4.60 5.97   1.75 2.01 4.02 2.34 7.20 0.69 3.10 3.92 11.71
 1.56 3.03 4.01 2.61 2.88   5.97 6.24 7.89 5.11 3.36 1.56 7.50 2.16  1.33
 1.42 2.76 2.17 3.41 3.47   3.15 4.08 2.29 3.95 5.42 1.77 2.80 9.69  3.95
```

```
 9.04 1.38 2.61 1.14 5.24   1.42 2.06 5.46 3.72 9.80 2.77 1.71 7.25   2.86
 5.15 2.94 3.00 1.90 4.61   3.64 7.54 1.85 2.50 0.95 1.14 1.85 3.97   6.06
 4.47 6.69 2.02 6.04 5.63   5.17 2.12 3.70 1.72 3.50 3.73 8.03 6.87   5.01
 1.07 5.17 4.97 2.99 2.45   5.82 5.50 5.34 4.65 4.73 2.91 4.75 1.45   4.27
 3.71 3.16 5.82 6.24
;

proc means data=MRData N min max mean std skew kurt maxdec=3;
run;

proc univariate data=MRData;
   histogram x / midpoints=(0 to 13);
   ods select Histogram;
run;
```

Figure 16.4 Summary Statistics

The MEANS Procedure

	Analysis Variable : x					
N	Minimum	Maximum	Mean	Std Dev	Skewness	Kurtosis
200	0.560	13.180	4.022	2.153	1.152	2.122

Figure 16.5 Histogram of Sample

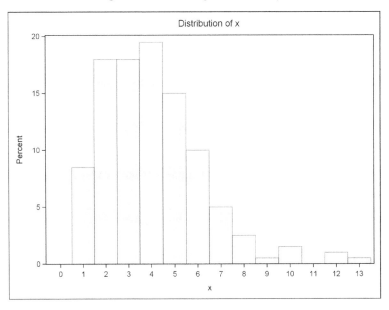

As shown in Figure 16.1, there are various ways to produce simulated samples that match the characteristics of the observed data. For each technique, you can plot the skewness and kurtosis of the simulated samples on the moment-ratio diagram. This enables you to visually assess the sampling variability of these statistics.

16.5.2 Plotting Bootstrap Estimates of Standard Errors

The following statements use the basic bootstrap technique to generate 100 samples (technically *re*samples) of the MRData data set. The estimates of skewness and kurtosis for each bootstrap sample are plotted on the moment-ratio diagram in Figure 16.6. Reference lines are added to show the skewness and kurtosis values of the observed data. The macro that creates the moment-ratio diagram is available from this book's Web site.

```
/* use SURVEYSELECT to generate bootstrap resamples */
proc surveyselect data=MRData out=BootSamp noprint
      seed=12345 method=urs rep=100 rate=1;
run;

proc means data=BootSamp noprint;
   by Replicate;
   freq NumberHits;
   var x;
   output out=MomentsBoot skew=Skew kurt=Kurt;
run;

title "Moment-Ratio Diagram";
title2 "100 Bootstrap Resamples, N=200";
%PlotMRDiagramRef(MomentsBoot, anno, Transparency=0.4);
```

Figure 16.6 Skewness and Kurtosis of Bootstrap Resamples

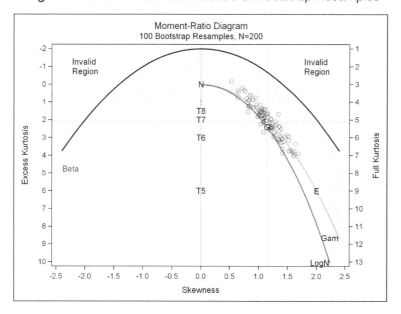

Vargo, Pasupathy, and Leemis (2010) suggest using the spread of the moment-ratio "cloud" to guide the selection of a model. A candidate distribution is one that is close to the center of the cloud. (You can also fit a bivariate normal predicton ellipse to the points of the cloud; see Vargo, Pasupathy, and Leemis (2010) for an example.)

For the current example, plausible candidates include the gamma distribution, the Gumbel distribution, and the lognormal families. The next section explores simulating the data with a fitted gamma distribution.

Exercise 16.3: Use the HISTOGRAM statement in PROC UNIVARIATE to estimate parameters for the candidate distributions and overlay the density distributions. Read the next section if you need help fitting distributions to the data.

16.6   Fitting a Gamma Distribution to Data

Suppose that you decide to use a gamma family to fit to the MRData data. This decision might be motivated by some domain-specific knowledge about the data, or it might be made by noticing that the sample moment-ratios in Figure 16.6 lie along the curve for the gamma family.

You can use the UNIVARIATE procedure to fit a gamma distribution to the data. The UNIVARIATE procedure uses a generalization of the gamma distribution that includes a threshold and scale parameter. The following statements set the threshold parameter to zero and fit the scale and shape parameters. The parameter estimates are shown in Figure 16.7. The INSET statement is used to display the skewness and kurtosis values on the histogram, which is shown in Figure 16.8.

```
proc univariate data=MRData;
   var x;
   histogram x / gamma(theta=0);
   inset skewness kurtosis / format=5.3;
   ods select Histogram ParameterEstimates;
run;
```

Figure 16.7 Parameter Estimates for a Gamma Distribution

The UNIVARIATE Procedure
Fitted Gamma Distribution for x

Parameters for Gamma Distribution		
Parameter	Symbol	Estimate
Threshold	Theta	0
Scale	Sigma	1.124407
Shape	Alpha	3.57664
Mean		4.0216
Std Dev		2.12648

Figure 16.7 shows the parameter estimates for the scale and shape parameters, which are computed by using maximum likelihood (ML) estimation. The mean and standard deviation of the fitted distribution are also displayed. Goodness-of-fit statistics are not shown but indicate that the fit is good. As shown in Table 16.1, the skewness of a gamma distribution with shape parameter $\alpha = 3.58$ is $2/\sqrt{3.58} = 1.06$, and the excess kurtosis is $6/3.58 = 1.68$. Notice that these ML estimates do not match the sample skewness and kurtosis, which is to be expected. (An alternative approach would be to find the value of α such that $(2/\sqrt{\alpha}, 6/\alpha)$ is closest to the sample skewness and kurtosis.)

Figure 16.8 Distribution of Data with Fitted Gamma Density

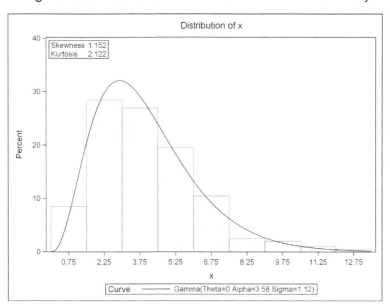

The model fitting is complete. You can now simulate from the fitted model.

An interesting application of the moment-ratio diagram is to assess the variability of the skewness and kurtosis of the simulated samples. Figure 16.6 shows the variability in those statistics for the original data as estimated by the bootstrap method. Does simulating from the fitted gamma distribution lead to a similar picture?

The following DATA step simulates the data from the fitted gamma model and computes the skewness and kurtosis for each sample. These values are overlaid on the moment-ratio diagram in Figure 16.9.

```
%let N = 200;                                      /* match the sample size */
%let NumSamples = 100;
data Gamma(keep=x SampleID);
call streaminit(12345);
do SampleID = 1 to &NumSamples;
   do i = 1 to &N;
      x = 1.12 * rand("Gamma", 3.58);
      output;
   end;
end;
run;

proc means data=Gamma noprint;
   by SampleID;
   var x;
   output out=MomentsGamma mean=Mean var=Var skew=Skew kurt=Kurt;
run;

title "Moment-Ratio Diagram";
title2 "&NumSamples Samples from Fitted Gamma, N=&N";
%PlotMRDiagramRef(MomentsGamma, anno, Transparency=0.4);
```

Figure 16.9 Skewness and Kurtosis of Simulated Samples: Gamma Model

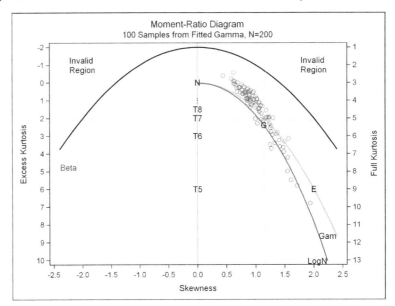

Figure 16.9 shows the skewness-kurtosis values for each sample of size 200. Again, reference lines show the skewness and kurtosis values of the observed data. Figure 16.9 looks similar to Figure 16.6, although the dispersion of the cloud for the gamma model appears to be greater than for the bootstrap cloud. The sample skewness ranges from 0.4 to 1.9, and the excess kurtosis ranges from −0.6 to 6.8.

Exercise 16.4: An advantage of a parametric model is that you can simulate different sample sizes. Rerun the simulation with $N = 10,000$ observations in each sample. Create a scatter plot of the sample skewness and kurtosis for each simulation. (You do not need to overlay the moment-ratio diagram.) What do you observe about the size of the cloud?

Exercise 16.5: Use PROC UNIVARIATE to fit a Gumbel distribution to the data. Simulate 100 random samples from the fitted Gumbel distribution, each containing 200 observations. (See Section 7.3 for how to simulate Gumbel data.) Create a scatter plot of the sample skewness and kurtosis for the simulated data and compare it to Figure 16.9.

16.7   The Johnson System

The previous section assumes that the gamma distribution is an appropriate model for the data. However, sometimes you simply do not know whether any "named" family is appropriate for a given set of data. In that case, you might turn to a flexible system of distributions that can be used to fit a wider variety of distributional shapes.

Johnson's system (Johnson 1949) consists of three families of distributions that are defined by normalizing transformations: the S_B distributions, the S_L distributions (which are the lognormal distributions), and the S_U distributions. Section 7.3.7 and Section 7.3.8 discuss the distributions and provide DATA steps that simulate data from them.

Together, these distributions (and the normal distribution) cover the range of valid moment ratios as described in Section 16.3. Figure 16.10 shows that the lognormal distribution (and its reflection) separates the moment-ratio diagram into two regions. Distributions in the upper region have relatively low kurtosis. Each (γ, κ) value in the upper region is achieved by one S_B distribution. The lower region corresponds to high kurtosis values. Each (γ, κ) value in the lower region is achieved by one S_U distribution.

Figure 16.10 Moment Ratios for the Johnson System of Distributions

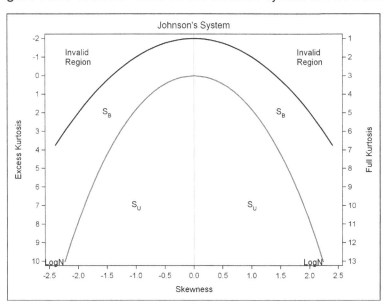

The sample moment ratios for the MRData data are in the S_B region. The UNIVARIATE procedure can use the first four sample moments to fit a family in the Johnson system. Be aware, however, that the variances of the estimates of the third and fourth moments are quite high, that the estimates are biased for small samples, and that the estimates are not robust to outliers (Slifker and Shapiro 1980). The following statements use the method of moments to fit the S_B family to the MRData data. Figure 16.11 shows the parameter estimates and the associated moments for the S_B distribution. The UNIVARIATE procedure does not provide goodness-of-fit statistics for the S_B family, but Figure 16.12 indicates that the model appears to fit the data well.

```
proc univariate data=MRData;
   var x;
   histogram x / sb(fitmethod=Moments theta=est sigma=est);
   ods output ParameterEstimates=PE;
   ods select Histogram ParameterEstimates;
run;
```

Figure 16.11 Parameter Estimates for a S_B Distribution

The UNIVARIATE Procedure
Fitted SB Distribution for x

Parameters for Johnson SB Distribution		
Parameter	Symbol	Estimate
Threshold	Theta	-0.87691
Scale	Sigma	50.63019
Shape	Delta	2.082419
Shape	Gamma	4.84161
Mean		4.021602
Std Dev		2.153048
Skewness		1.151726
Kurtosis		2.121984
Mode		2.888916

Figure 16.12 Distribution of Data with Fitted S_B Density

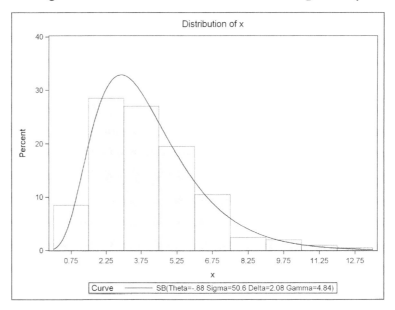

The model fitting is complete. You can now simulate from the fitted model as described in Section 7.3.7. You can use the moment-ratio diagram to assess the variability of the skewness and kurtosis of the simulated samples. Figure 16.13 displays the cloud of the sample skewness and kurtosis values. The cloud is similar to Figure 16.9 and includes reference lines that show the skewness and kurtosis of the observed data.

Figure 16.13 Skewness and Kurtosis of Simulated Samples: Johnson S_B Distribution

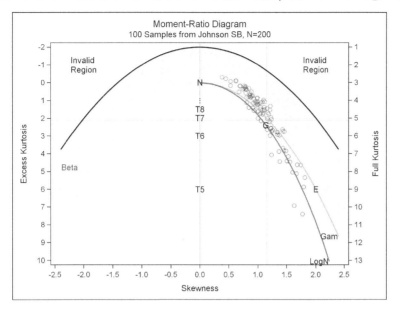

16.8   Fleishman's Method

Fleishman (1978) uses a cubic transformation of a standard normal variable to construct a distribution with given moments. The transformation starts with a standard normal random variable, Z, and generates a new random variable defined by the polynomial

$$Y = c_0 + c_1 Z + c_2 Z^2 + c_3 Z^3$$

Given specific values of skewness and kurtosis, you can often (but not always) find coefficients so that Y has the given skewness and kurtosis. Fleishman (1978) tabulated coefficient values (always setting $c_0 = -c_2$) for selected values of skewness and kurtosis. Alternatively, many researchers (for example, Fan et al. (2002)) use root-finding techniques to solve numerically for the coefficients that give specified values of skewness and kurtosis.

For years there were two main objections to using Fleishman's method (Tadikamalla 1980):

- For a given value of skewness, γ, there are distributions with (full) kurtosis κ for all $\kappa \geq 1 + \gamma^2$. However, the distributions that are generated by using Fleishman's transformation cannot achieve certain low values of kurtosis. In particular, Fleishman distributions are bounded by the equation $\kappa \geq 1.8683 + 1.58837\gamma^2$.

- The distribution of Y was not known in terms of an analytic expression. Although it was possible to generate random samples from Y, quantiles, modes, and other important features were not available.

Figure 16.14 shows the region $\kappa \geq 1.8683 + 1.58837\gamma^2$ that can be achieved by cubic transformations of a standard normal variate. Any (γ, κ) value below the dashed curve can be generated by Fleishman's transformation.

Figure 16.14 Moment-Ratio Diagram with Fleishman Region

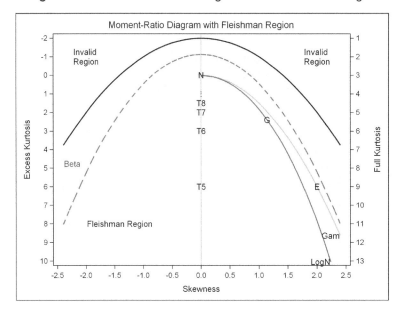

Today, however, these objections are less troublesome thanks to recent research that was primarily performed by Todd Headricks and his colleagues.

- Fleishman's cubic transformation method has been extended to higher-order polynomials (Headrick 2002, 2010) and is now the simplest example of the general "power transformation" method of generating nonnormal distributions. Using higher-order polynomials decreases (but does not eliminate) the region of unobtainable (γ, κ) values at the cost of solving more complicated equations.

- Headrick and Kowalchuk (2007) published a description of the PDF and CDF for the power transformation method.

Fleishman's method is often used in practice because it is fast, relatively easy, and it extends to multivariate correlated data (Vale and Maurelli 1983). Headrick (2010) is the authoritative reference on the properties of Fleishman's transformation.

This book's Web site contains a series of SAS/IML functions that you can use to find the coefficients of Fleishman's cubic transformation for any obtainable skewness and kurtosis values. You can download and store the functions in a SAS/IML library, and use the LOAD statement to read the modules into the active PROC IML session.

There are three important SAS/IML functions for using the Fleishman method:

- The MOMENTS module, which computes the sample mean, variance, skewness, and kurtosis for each column of an $N \times p$ matrix.

- The FITFLEISHMAN module, which returns a vector of coefficients $\{c_0, c_1, c_2, c_3\}$ that transforms the standard normal distribution into a distribution (the *Fleishman distribution*) that has the same moment ratios as the data.

- The RANDFLEISHMAN module, which generates an $N \times p$ matrix of random variates from the Fleishman distribution, given the coefficients $\{c_0, c_1, c_2, c_3\}$.

The following SAS/IML program reads the MRData data and finds the cubic coefficients for the Fleishman transformation. As in previous sections, you can simulate 100 samples from the Fleishman distribution that has the same skewness and kurtosis as the data and plot the moment-ratios, as shown in Figure 16.15.

```
/* Define and store the Fleishman modules */
%include "C:\<path>\RandFleishman.sas";

/* Use Fleishman's cubic transformation to model data */
proc iml;
load module=_all_;                         /* load Fleishman modules */
use MRData;  read all var {x};  close MRData;

c = FitFleishman(x);      /* fit model to data; obtain coefficients */

/* Simulate 100 samples, each with nrow(x) observations */
call randseed(12345);
Y = RandFleishman(nrow(x), 100, c);
Y = mean(x) + std(x) * Y;                 /* translate and scale Y  */

varNames = {"Mean" "Var" "Skew" "Kurt"};
m = T( Moments(Y) );
create MomentsF from m[c=varNames];  append from m;  close MomentsF;
quit;

title "Moment-Ratio Diagram";
title2 "&NumSamples Samples from Fleishman Distribution, N=&N";
%PlotMRDiagramRef(MomentsF, Fanno, Transparency=0.4);
```

Figure 16.15 shows the sample skewness and kurtosis of 100 samples that are generated by using a cubic transformation of normal variates. Reference lines show the skewness and kurtosis values of the observed data. Notice that all moment ratios are inside the Fleishman region, which is indicated by the dashed parabola. The variability of the moment-ratios appears to be similar to that shown in Figure 16.13 for the Johnson family.

When matching moments to data, there are an infinite number of distributions that give the same set of moments. The Johnson family and the Fleishman transformation are two useful techniques to simulate data when the first four moments are specified. Because four moments do not completely characterize a distribution, you should always plot a histogram of a simulated sample as a check that the model is appropriate for the data.

Figure 16.15 Skewness and Kurtosis of Simulated Samples: Fleishman Distribution

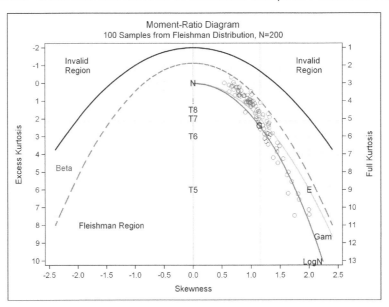

16.9 Comparing Simulations and Choosing Models

As shown in the previous sections, the moment-ratio diagram enables you to compare simulations from different models. When you present the results of a simulation study, be sure to explain the details of the study so that others can reproduce, extend, or critique the methods used.

Section 16.6 fits a model from the gamma distribution and simulates data from the fitted model. This is the usual approach: You choose a parametric family and use the data to estimate the parameters. As shown in Section 16.5, you can use the moment-ratio diagram to help choose candidates for modeling.

Section 16.7 and Section 16.8 uses a moment-matching approach to fit the data. The Johnson system and the Fleishman transformation are flexible tools that can fit a wide range of distribution shapes.

The Johnson system can be used to fit any feasible pair of skewness and kurtosis values. However, it is not always easy to fit a Johnson distribution to data. The Fleishman transformation is not quite as flexible, but has the advantage of being a single method rather than a system of three separate distributions. You can use the programs from this book's Web site to generate samples from distributions that have specified moments.

No matter what method you use to simulate data, you can use the moment-ratio diagram to assess the variability of the sample skewness and kurtosis.

16.10   The Moment-Ratio Diagram As a Tool for Designing Simulation Studies

In addition to its usefulness as a modeling tool, moment matching is a valuable technique that you can use to systematically simulate data from a wide range of distributional shapes.

Imagine that you want to use simulation to explore how a statistic or statistical technique behaves for nonnormal data. A well-designed simulation study should choose distributions that are representative of the variety of nonnormal distributions. Instead of asking, "How is the statistic affected by nonnormal data?" you can ask instead, "How is the statistic affected by the skewness and kurtosis of the underlying distribution?" The new question leads to the following approach, which simulates data from distributions with a specified range of skewness and kurtosis:

1. Define a regularly spaced grid of skewness and kurtosis values. In SAS/IML software, you can use the EXPAND2DGRID function, which is defined in Appendix A.

2. For each feasible pair of skewness and kurtosis, use Fleishman's method to construct a distribution with the given skewness and kurtosis.

3. Simulate data from the constructed distribution. Apply the statistical method to the samples and compute a measure of the performance of the method.

4. Construct a graph that summarizes how the measure depends on the skewness and kurtosis.

To be specific, suppose that you are interested in the coverage probability of the confidence interval for the mean given by the formula $[\bar{x} - t_{0.975,n-1}s/\sqrt{n}, \ \bar{x} + t_{0.975,n-1}s/\sqrt{n}]$. Here $\bar{x}$ in the sample mean, s is the sample standard deviation, and $t_{1-\alpha/2,n-1}$ is the $(1 - \alpha/2)$ percentile of the t distribution with $n - 1$ degrees of freedom. As shown in Section 5.2, for a random sample that is drawn from a normal distribution, this interval is an exact 95% confidence interval. However, for a sample drawn from a nonnormal distribution, the coverage probability is often different from 95%. The goal of this section is to study the effect of skewness and kurtosis on coverage probability.

You can use PROC MEANS to compute the confidence intervals, and then use a DATA step and PROC FREQ to estimate the coverage probability, as shown in Section 5.2.1. Alternatively, you can write a SAS/IML module to compute the confidence intervals and the coverage estimates, as follows:

```
proc iml;
/* Assume X is an N x p matrix that contains p samples of size N.
   Return a 2 x p matrix where the first row is xbar - t*s/sqrt(N) and
   the second row is xbar + t*s/sqrt(N), where s=Std(X) and t is the
   1-alpha/2 quantile of the t distribution with N-1 degrees of freedom */
start NormalCI(X, alpha);
   N = nrow(X);
   xbar = mean(X);
   t = quantile("t", 1-alpha/2, N-1);
   dx = t*std(X)/sqrt(N);
   return ( (xbar - dx) // (xbar + dx) );
finish;
```

You can use the NORMALCI function and the Fleishman functions that are available on this book's Web site to simulate data from distributions with a specified range of skewness and kurtosis. For this simulation, the pairs of skewness and kurtosis are generated on an equally spaced grid, which is shown in Figure 16.16. The skewness is chosen in the interval [0, 2.4], and the kurtosis is chosen to be less than 10.

```
call randseed(12345);
load module=_all_;                              /* load Fleishman modules */

/* 1. Create equally spaced grid in (skewness, kurtosis) space:
      {0, 0.2, 0.4,..., 2.4} x {-2, -1.5, -1,..., 10} */
sk = Expand2DGrid( do(0,2.4,0.2), do(-2,10,0.5) );
skew = sk[,1]; kurt = sk[,2];
/* 1a. Discard invalid pairs */
idx =  loc(kurt > (-1.2264489 + 1.6410373# skew##2));
sk = sk[idx, ];                                 /* keep these values    */
skew = sk[,1]; kurt = sk[,2];

/* for each (skew, kurt), use simul to estimate coverage prob for CI */
N = 25;
NumSamples = 1e4;
Prob = j(nrow(sk), 1, .);
do i = 1 to nrow(sk);
   c = FitFleishmanFromSK(skew[i], kurt[i]); /* find Fleishman coefs */
   X = RandFleishman(N, NumSamples, c);      /* generate samples     */
   CI = NormalCI(X, 0.05);                   /* compute normal CI    */
   Prob[i] = ( CI[1,]<0 & CI[2,]>0 )[:];     /* mean of indicator var*/
end;
```

Every sample in this simulation study is drawn from a population with zero mean. The **Prob** vector contains the proportion of simulated samples for which the confidence interval includes zero. This proportion estimates the coverage probability of the interval. For normal data, this proportion is about 95%. For distributions with larger skewness, the coverage probability will be less.

A plot of the coverage probability as a function of the skewness and kurtosis indicates that a quadratic smoother is a good fit. The following statements write the results of the simulation study to a data set and use the RSREG procedure to fit a quadratic surface to the probability estimates as a function of the skewness and kurtosis. A contour plot of the fitted surface is shown in Figure 16.16.

```
create MRGrid var {"Skew" "Kurt" "Prob"};  append;  close;
quit;

proc rsreg data=MRGrid plots=Contour;
   model Prob = Kurt Skew;
   ods select Contour;
run;
```

Figure 16.16 Smoothed Results of Simulation Study

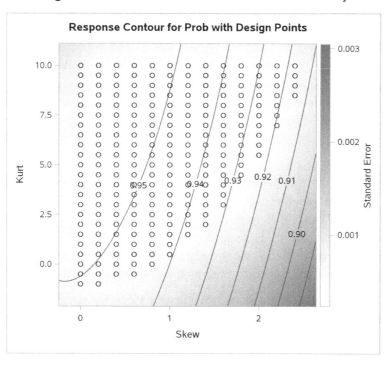

The graph summarizes the results of this simulation study, which includes 214 pairs of skewness-kurtosis values. (Unfortunately, the kurtosis axis points up in the figure.) For distributions with small values of skewness, the coverage probability is about 95%. For larger values of skewness, the coverage probability decreases. The contours in Figure 16.16 show that the kurtosis is important, but kurtosis is not as important as the skewness. The relative flatness of the fitted surface indicates that you can probably use a much coarser grid and draw the same conclusions.

For many simulation studies that include parameters, it is a good idea to first run a small-scale simulation on a coarse grid or on a set of parameter values that are produced by an experimental design. The small-scale results can be analyzed to determine areas of the parameter space for which the response function (in this case, the coverage probability) is changing quickly and areas where it is essentially constant. Areas in which the response function is flat do not require many parameter values. You can refine the grid of parameter values in those areas for which the response variable changes quickly. By using this strategy, you can run a few simulations for "uninteresting" parameter values and more simulations for "interesting" values.

Exercise 16.6: Rerun the simulation study, but include samples from distributions with negative skewness by creating a grid of values as follows:

```
sk = Expand2DGrid( do(-2.4,2.4,0.2), do(-2,10,0.5) );
```

Do the results depend on the sign of the skewness?

16.11   Extensions to Multivariate Data

In Vale and Maurelli (1983), the authors mention that of all the methods studied by Tadikamalla (1980), "Fleishman's procedure was the easiest to implement and executed most quickly." Furthermore, "Fleishman's procedure has an advantage over the other procedures in that it can easily be extended to generate multivariate random numbers with specified intercorrelations and univariate [moments]."

This section describes Vale and Maurelli's algorithm for simulating multivariate nonnormal data. As discussed in Section 8.9, a technique for generating multivariate correlated data is to start with uncorrelated normal variables and transform the variables according to a square root of the desired correlation matrix. Specifically, if V is a matrix and $V = UDU'$ is the spectral decomposition of V, then the matrix $F = D^{1/2}U'$ is a square root matrix of V because $F'F = V$. When V is a correlation matrix, uncorrelated normal variables Z_i are transformed by F into correlated multivariate normal data with correlations given by the elements of V.

Vale and Maurelli note that you can construct the matrix V so that when you use the Fleishman transformation on each variable, the marginal distributions have the specified skewness, kurtosis, and correlations. Specifically, the Vale-Maurelli algorithm is as follows:

1. Specify the skewness and kurtosis values for the marginal distributions, and compute the Fleishman coefficients for each.

2. Specify the correlations, R, between variables, and compute the intermediate correlation matrix $V = V(R)$ by solving for the roots of a cubic equation that involves the Fleishman coefficients.

3. Use the spectral decomposition, $V = UDU'$, to compute a square root matrix, $F = D^{1/2}U'$.

4. Simulate uncorrelated variables $Z = (Z_1, Z_2, \ldots, Z_p)$.

5. Use the F and Z matrices to form correlated variables $Y = (Y_1, Y_2, \ldots, Y_p)$.

6. Apply the Fleishman cubic transformation to form $X_i = \Sigma_{j=0}^{3} c_j Y_i^j$.

The final data $X = (X_1, X_2, \ldots, X_p)$ are from a distribution with the given correlations and where the marginal distributions have the specified skewness and kurtosis.

This book's Web site contains SAS/IML functions that implement the Vale-Maurelli algorithm. In the following program, two SAS/IML functions are loaded from storage:

- the VMTARGETCORR function, which returns the matrix V, given the correlation matrix R and the values of the skewness and kurtosis for the marginal distributions

- the RANDVALEMAURELLI function, which generates an $N \times p$ matrix of random variates according to the Vale-Maurelli algorithm

To test the algorithm, you can sample from a multivariate distribution that has three variables. The following example chooses marginal distributions that have the same skewness and excess kurtosis as the Gamma(4), exponential, and t_5 distributions. The skewness and excess kurtosis are computed by using the formulas in Table 16.1. The correlation matrix is specified arbitrarily as a 3×3 symmetric matrix where $R_{12} = 0.3$, $R_{13} = -0.4$, and $R_{23} = 0.5$. The following statements compute C, which is the intermediate matrix of pairwise correlations. Figure 16.17 shows the matrix of pairwise correlations that results from step 2 of the Vale-Maurelli algorithm. It is this matrix (which is not guaranteed to be positive definite) that is used to induce correlations in step 5 of the algorithm.

```
/* Define and store the Vale-Maurelli modules */
%include "C:\<path>\RandFleishman.sas";

proc iml;
load module=_all_;       /* load Fleishman and Vale-Maurelli modules */

/* Find target correlation for given skew, kurtosis, and corr.
   The (skew,kurt) correspond to Gamma(4), Exp(1), and t5.    */
skew = {1    2 0};
kurt = {1.5 6 6};
R = {1.0   0.3 -0.4,
     0.3   1.0  0.5,
    -0.4   0.5  1.0 };

V = VMTargetCorr(R, skew, kurt);
print V[format=6.4];
```

Figure 16.17 Intermediate Matrix of Pairwise Correlations

V		
1.0000	0.3318	-.4243
0.3318	1.0000	0.5742
-.4243	0.5742	1.0000

Notice that there is nothing random about the output from the VMTARGETCORR function. Randomness occurs only in step 4 of the algorithm.

You can obtain samples by calling the RANDVALEMAURELLI function, which returns an $N \times 3$ matrix for this example, as follows:

```
/* generate samples from Vale-Maurelli distribution */
N = 10000;
call randseed(54321);
X = RandValeMaurelli(N, R, skew, kurt);
```

To verify that the simulated data in **x** have the specified properties, you can use PROC MEANS and PROC CORR to compute the sample moments and to visualize the multivariate distribution. The results are shown in Figure 16.18. Because estimates of the skewness and kurtosis are highly variable, 10,000 observations are used. However, to avoid overplotting only 1,000 observations are used to visualize the distribution of the simulated data, as follows:

```
create VM from X[c=("x1":"x3")]; append from X; close VM;
quit;

proc means data=VM N Mean Var Skew Kurt;
run;

proc corr data=VM(obs=1000) noprob plots(maxpoints=NONE)=matrix(histogram);
ods select PearsonCorr MatrixPlot;
run;
```

Figure 16.18 Descriptive Statistics of Simulated Data

The MEANS Procedure

Variable	N	Mean	Variance	Skewness	Kurtosis
x1	10000	0.0018588	1.0196136	1.0127822	1.4446137
x2	10000	0.0135446	1.0334302	2.0696081	6.4912644
x3	10000	0.0086906	0.9858476	0.0076107	4.5343643

The CORR Procedure

Pearson Correlation Coefficients, N = 1000			
	x1	x2	x3
x1	1.00000	0.36912	-0.40994
x2	0.36912	1.00000	0.43593
x3	-0.40994	0.43593	1.00000

Figure 16.18 shows that the sample means are close to zero and the sample variances are close to unity. Taking into account sampling variation, the skewness and kurtosis are close to the specified values as are the correlations between variables.

Figure 16.19 shows the distribution of the simulated data. The Vale-Maurelli algorithm has generated nonnormal data with specified moments and with specified correlations.

Figure 16.19 Simulated Data by Using the Vale-Maurelli Algorithm

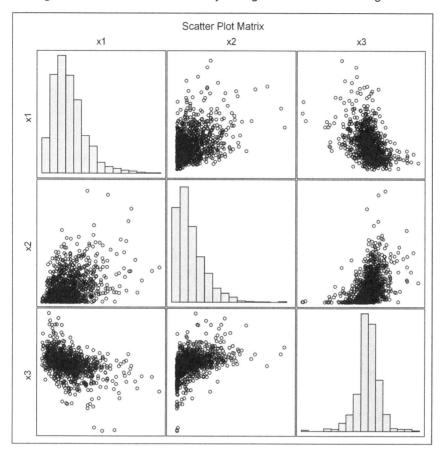

The Vale-Maurelli algorithm is simple to implement and usually works well in practice. Headrick and Sawilowsky (1999) note that the Vale-Maurelli algorithm can break down when the conditional distributions are highly skewed or heavy-tailed, and when the sample sizes are small. They propose an algorithm that is algebraically more complex but which seems to handle these more extreme situations.

16.12 References

Bowman, K. O. and Shenton, L. R. (1983), "Johnson's System of Distributions," in S. Kotz, N. L. Johnson, and C. B. Read, eds., *Encyclopedia of Statistical Sciences*, volume 4, 303–314, New York: John Wiley & Sons.

Burr, I. W. (1942), "Cumulative Frequency Functions," *Annals of Mathematical Statistics*, 13, 215–232.

Fan, X., Felsoványi, A., Sivo, S. A., and Keenan, S. C. (2002), *SAS for Monte Carlo Studies: A Guide for Quantitative Researchers*, Cary, NC: SAS Institute Inc.

Fleishman, A. (1978), "A Method for Simulating Non-normal Distributions," *Psychometrika*, 43, 521–532.

Headrick, T. C. (2002), "Fast Fifth-Order Polynomial Transforms for Generating Univariate and Multivariate Nonnormal Distributions," *Computational Statistics and Data Analysis*, 40, 685–711.

Headrick, T. C. (2010), *Statistical Simulation: Power Method Polynomials and Other Transformations*, Boca Raton, FL: Chapman & Hall/CRC.

Headrick, T. C. and Kowalchuk, R. K. (2007), "The Power Method Transformation: Its Probability Density Function, Distribution Function, and Its Further Use for Fitting Data," *Journal of Statistical Computation and Simulation*, 77, 229–249.

Headrick, T. C. and Sawilowsky, S. (1999), "Simulating Correlated Multivariate Nonnormal Distributions: Extending the Fleishman Power Method," *Psychometrika*, 64, 25–35.

Johnson, N. L. (1949), "Systems of Frequency Curves Generated by Methods of Translation," *Biometrika*, 36, 149–176.

Johnson, N. L., Kotz, S., and Balakrishnan, N. (1994), *Continuous Univariate Distributions*, volume 1, 2nd Edition, New York: John Wiley & Sons.

Kendall, M. G. and Stuart, A. (1977), *The Advanced Theory of Statistics*, volume 1, 4th Edition, New York: Macmillan.

Ord, J. K. (2005), "Pearson System of Distributions," in S. Kotz, N. Balakrishnan, C. B. Read, B. Vidakovic, and N. L. Johnson, eds., *Encyclopedia of Statistical Sciences*, volume 9, 2nd Edition, 6036–6040, New York: John Wiley & Sons.

Rodriguez, R. N. (2005), "Burr Distributions," in S. Kotz, N. Balakrishnan, C. B. Read, B. Vidakovic, and N. L. Johnson, eds., *Encyclopedia of Statistical Sciences*, volume 1, 2nd Edition, 678–683, New York: John Wiley & Sons.

Slifker, J. F. and Shapiro, S. S. (1980), "The Johnson System: Selection and Parameter Estimation," *Technometrics*, 22, 239–246.

Tadikamalla, P. (1980), "On Simulating Non-normal Distributions," *Psychometrika*, 45, 273–279.

Vale, C. and Maurelli, V. (1983), "Simulating Multivariate Nonnormal Distributions," *Psychometrika*, 48, 465–471.

Vargo, E., Pasupathy, R., and Leemis, L. M. (2010), "Moment-Ratio Diagrams for Univariate Distributions," *Journal of Quality Technology*, 42, 1–11.

Part V

Appendix

Appendix A
A SAS/IML Primer

Contents

A.1 Overview of the SAS/IML Language

The SAS/IML language is a high-level matrix language that enables SAS users to develop algorithms and compute statistics that are not built into any SAS procedure. The language contains hundreds of built-in functions for statistics, data analysis, and matrix computations, and enables you to call hundreds of DATA step functions. You can write your own functions to extend the language.

If you are serious about simulating data (especially multivariate data), you should take the time to learn the SAS/IML language. The following resources can help you get started:

- Read chapters 1–4 and 13–15 of *Statistical Programming with SAS/IML Software* (Wicklin 2010).

- Subscribe to The DO Loop blog, which is a statistical programming blog that is located at the URL `blogs.sas.com/content/iml`.

- Ask questions at the SAS/IML Community, which is located at `communities.sas.com/community/support-communities`.

- Read the first few chapters of the *SAS/IML User's Guide*.

A.2 SAS/IML Functions That Are Used in This Book

It is assumed that the reader is familiar with

- Basic DATA step functions such as SQRT, CEIL, and EXP. When used in SAS/IML software, these functions operate on every element of a matrix.

- Statistical functions such as PDF, CDF, and QUANTILE (see Section 3.2). These functions also act on every element of a matrix. In certain cases, you can pass in vectors of parameters to these functions.

- Control statements such as IF-THEN/ELSE and the iterative DO statement.

This section describes SAS/IML functions and subroutines that are used in this book. The definitions are taken from the *SAS/IML User's Guide*. Note: The functions marked with an asterisk (*) were introduced in SAS/IML 12.1, which is distributed as part of the second maintenance release of SAS 9.3.

ALL function	checks for all nonzero elements
ANY function	checks for any nonzero elements
BLOCK function	forms block-diagonal matrices
CHOOSE function	evaluates a logical matrix and returns values based on whether each element is true or false
COLVEC function	reshapes a matrix into a column vector
CORR function	computes correlation statistics
COUNTN function	counts the number of nonmissing values
COV function	computes a sample variance-covariance matrix
CUPROD function	computes cumulative products
CUSUM function	computes cumulative sums
DIAG function	creates a diagonal matrix
DISTANCE function*	computes pairwise distances between rows of a matrix
DO function	produces an arithmetic series
EIGEN call	computes eigenvalues and eigenvectors
EIGVAL function	computes eigenvalues
FINISH statement	denotes the end of a module

FREE statement	frees the memory associated with a matrix
FROOT function*	numerically finds zeros of a univariate function
I function	creates an identity matrix
INV function	computes the inverse
J function	creates a matrix of identical values
LOAD statement	loads modules and matrices from library storage
LOC function	finds indices for the nonzero elements of a matrix
MAX function	finds the maximum value of a matrix
MEAN function	computes sample means
MEDIAN function	computes sample medians
MIN function	finds the smallest element of a matrix
NCOL function	finds the number of columns of a matrix
NROW function	finds the number of rows of a matrix
POLYROOT function	finds zeros of a real polynomial
PROD function	computes products
QNTL call	computes sample quantiles (percentiles)
RANDGEN call	generates random numbers from specified distributions
RANDMULTINOMIAL function	generates a random sample from a multinomial distribution
RANDMVT function	generates a random sample from a multivariate Student's t distribution
RANDNORMAL function	generates a random sample from a multivariate normal distribution
RANDSEED call	initializes seed for subsequent RANDGEN calls
RANDWISHART function	generates a random sample from a Wishart distribution
RANK function	ranks elements of a matrix, breaking ties arbitrarily
REPEAT function	creates a matrix of repeated values
RETURN statement	returns from a module
ROOT function	performs the Cholesky decomposition of a matrix
ROWVEC function	reshapes a matrix into a row vector
SAMPLE function*	generates a random sample of a finite set
SETDIF function	compares elements of two matrices
SHAPE function	reshapes a matrix
SOLVE function	solves a system of linear equations
SORT call	sorts a matrix by specified columns
SQRVECH function	converts a symmetric matrix that is stored columnwise to a square matrix
SSQ function	computes the sum of squares of all elements
START statement	defines a module

STD function	computes a sample standard deviation
STOP statement	stops execution of statements
STORE statement	stores matrices and modules in a library
SUM function	computes sums
T function	transposes a matrix
TABULATE call	counts the number of unique values in a vector
TOEPLITZ function	generates a Toeplitz or block-Toeplitz matrix
TRISOLV function	solves linear systems with triangular matrices
UNION function	performs unions of sets
UNIQUE function	sorts and removes duplicates
VAR function	computes a sample variance
VECDIAG function	extracts the matrix diagonal into a vector
XSECT function	intersects sets

A.3   The PRINT Statement

The PRINT statement displays the data in one or more SAS/IML variables. The PRINT statement supports four options that control the output:

PRINT *x[COLNAME= ROWNAME= FORMAT= LABEL=]* ;

COLNAME=c
    specifies a character matrix to be used for the column heading of the matrix

ROWNAME=r
    specifies a character matrix to be used for the row heading of the matrix

FORMAT=$format$
    specifies a valid SAS or user-defined format to use to print the values of the matrix

LABEL=$label$
    specifies the character string to use as a label for the matrix

A.4   Subscript Reduction Operators in SAS/IML Software

One way to avoid writing unnecessary loops is to take full advantage of the subscript reduction operators for matrices. These operators enable you to perform common statistical operations (such as sums, means, and sums of squares) on the rows or the columns of a matrix. A common use of subscript reduction operators is to compute the marginal frequencies in a two-way frequency table.

The following table summarizes the subscript reduction operators for matrices and specifies an equivalent way to perform the operation that uses function calls.

Table A.1   Subscript Reduction Operators for Matrices

Operator	Action	Equivalent Function
+	Addition	sum(x)
#	Multiplication	prod(x)
><	Minimum	min(x)
<>	Maximum	max(x)
>:<	Index of minimum	loc(x=min(x))[1]
<:>	Index of maximum	loc(x=max(x))[1]
:	Mean	mean(x)
##	Sum of squares	ssq(x)

For example, the expression `x[+, ]` uses the `'+'` subscript operator to "reduce" the matrix by summing the elements of each row for all columns. (Recall that not specifying a column in the second subscript is equivalent to specifying all columns.) The expression `x[:, ]` uses the `':'` subscript operator to compute the mean for each column. Row sums and means are computed similarly. The subscript reduction operators correctly handle missing values.

A.5   Reading Data from SAS Data Sets

You can read each variable in a SAS data set into a SAS/IML vector, or you can read several variables into a SAS/IML matrix, where each column of the matrix corresponds to a variable. This section discusses both of these techniques.

A.5.1   Reading Data into SAS/IML Vectors

You can read data from a SAS data set by using the USE and READ statements. You can read variables into individual vectors by specifying a character matrix that contains the names of the variables that you want to read. The READ statement creates column vectors with those same names, as shown in the following statements:

```
proc iml;
/* read variables from a SAS data set into vectors */
varNames = {"Name" "Age" "Height"};
use Sashelp.Class(OBS=3);   /* open data set for reading              */
read all var varNames;      /* create three vectors: Name,...,Height */
close Sashelp.Class;        /* close the data set                    */
print Name Age Height;
```

Figure A.1   First Three Observations Read from a SAS Data Set

Name	Age	Height
Alfred	14	69
Alice	13	56.5
Barbara	13	65.3

A.5.2   Creating Matrices from SAS Data Sets

You can also read a set of variables into a matrix (assuming that the variables are either all numeric or all character) by using the INTO clause on the READ statement. The following statements illustrate this approach:

```
/* read variables from a SAS data set into a matrix */
varNames = {"Age" "Height" "Weight"};
use Sashelp.Class(OBS=3);
read all var varNames into m;    /* create matrix with three columns */
close Sashelp.Class;
print m[colname=VarNames];
```

Figure A.2 First Three Rows of a Matrix

m		
Age	**Height**	**Weight**
14	69	112.5
13	56.5	84
13	65.3	98

You can read only the numeric variable in a data set by specifying the _NUM_ keyword on the READ statement:

```
/* read all numeric variables from a SAS data set into a matrix */
use Sashelp.Class;
read all var _NUM_ into y[colname=NumericNames];
close Sashelp.Class;
print NumericNames;
```

Figure A.3 The Names of the Numeric Variables Read into a Matrix

NumericNames		
Age	Height	Weight

The matrix **NumericNames** contains the names of the numeric variables that were read; the columns of matrix **y** contain the data for those variables.

A.6   Writing Data to SAS Data Sets

You can write data in SAS/IML vectors to variables in a SAS data set, or you can create a data set from a SAS/IML matrix, where each column of the matrix corresponds to a variable.

A.6.1   Creating SAS Data Sets from Vectors

You can use the CREATE and APPEND statements to write a SAS data set from vectors or matrices. The following statements create a data set called OutData in the Work library:

```
/* create SAS data set from vectors */
x = T(1:10);                    /* {1,2,3,...,10}             */
y = T(10:1);                    /* {10,9,8,...,1}             */
create OutData var {x y};       /* create Work.OutData for writing */
append;                         /* write data in x and y      */
close OutData;                  /* close the data set         */
```

The CREATE statement opens Work.OutData for writing. The variables x and y are created; the type of the variables (numeric or character) is determined by the type of the SAS/IML vectors of the same name. The APPEND statement writes the values of the vectors listed on the VAR clause of the CREATE statement. The CLOSE statement closes the data set.

Row vectors and matrices are written to data sets as if they were column vectors. You can write character vectors as well as numeric vectors.

A.6.2   Creating SAS Data Sets from Matrices

To create a data set from a matrix of values, use the FROM clause on the CREATE and APPEND statements. If you do not explicitly specify names for the data set variables, the default names are COL1, COL2, and so forth. You can explicitly specify names for the data set variables by using the COLNAME= option in the FROM clause, as shown in the following statements:

```
/* create SAS data set from a matrix */
z = x || y;                          /* horizontal concatenation      */
create OutData2 from x[colname={"Count" "Value"}];
append from x;
close OutData2;
```

A.7   Creating an ID Vector

You can use the REPEAT and SHAPE (or COLVEC) functions to generate an ID variable as in Section 4.5.2.

For example, suppose that you have three patients in a study. Some measurement (for example, their weight) is taken every week for two weeks. You can order the data according to patient ID or according to time.

If you order the data by patient ID, then you can use the following statements to generate a categorical variable that identifies each observation:

```
proc iml;
N = 2;                                  /* size of each sample    */
NumSamples = 3;                         /* number of samples      */
ID = repeat( T(1:NumSamples), 1, N);    /* {1  1,
                                            2  2,
                                            3  3} */

SampleID = colvec(ID);                  /* convert to long vector */
```

The syntax **REPEAT(x, r, c)** stacks the values of the **x** matrix r times in the vertical direction and c times in the horizontal direction. The COLVEC function stacks the values (in row-major order) into a column vector.

If you order the data by time, then you can use the following statements to create an ID variable:

```
ID = repeat(1:NumSamples, 1, N);        /* {1  2  3  1  2  3  ... */
ReplID = colvec(ID);                     /* convert to long vector */
print SampleID ReplID;
```

Figure A.4 Two Ways to Construct an ID Variable

SampleID	ReplID
1	1
1	2
2	3
2	1
3	2
3	3

A.8   Creating a Grid of Values

It is useful to generate all pairwise combinations of elements in two vectors. For example, if **x** = {0, 1, 2} and **y** = {-1, 0, 1}, then the grid of pairwise values contains nine values as shown in Figure A.5.

```
proc iml;
/* Return ordered pairs on a regular grid of points.
   Return value is an (Nx*Ny x 2) matrix */
start Expand2DGrid( _x, _y );
   x  = colvec(_x); y  = colvec(_y);
   Nx = nrow(x);     Ny = nrow(y);
   x = repeat(x, Ny);
   y = shape( repeat(y, 1, Nx), 0, 1 );
   return ( x || y );
finish;

/* test the module */
x = {0,1,2};   y = {-1,0,1};
g = Expand2DGrid(x,y);
print g;
```

Figure A.5 A Grid of Values

g	
0	-1
1	-1
2	-1
0	0
1	0
2	0
0	1
1	1
2	1

A.9   Modules That Replicate SAS/IML Functions

Some function that are used in this book were introduced in SAS/IML 9.3 or SAS/IML 12.1. If you are using an earlier version of SAS/IML software, this section presents SAS/IML modules that reproduce the primary functionality of the functions.

A.9.1   The DISTANCE Function

The DISTANCE function computes pairwise distances between rows of a matrix and is used in Section 14.6. The EUCLIDEANDISTANCE and PAIRWISEDIST modules implement some of the functionality.

```
proc iml;
/* compute Euclidean distance between points in x and points in y.
   x is a p x d matrix, where each row is a point in d dimensions.
   y is a q x d matrix.
   The function returns the p x q matrix of distances, D, such that
   D[i,j] is the distance between x[i,] and y[j,]. */
start PairwiseDist(x, y);
   if ncol(x)^=ncol(y) then return (.);          /* Error              */
   p = nrow(x);   q = nrow(y);
   idx = T(repeat(1:p, q));                       /* index matrix for x */
   jdx = shape(repeat(1:q, p), p);                /* index matrix for y */
   diff = abs(X[idx,] - Y[jdx,]);
   return( shape( sqrt(diff[,##]), p ) );
finish;
```

```
/* compute Euclidean distance between points in x.
   x is a pxd matrix, where each row is a point in d dimensions. */
start EuclideanDistance(x);    /* in place of 12.1 DISTANCE function */
   y=x;
   return( PairwiseDist(x,y) );
finish;

x = { 1  0,
      1  1,
     -1 -1};
y = { 0  0,
     -1  0};
P = PairwiseDist(x,y);          /* not printed */
D = EuclideanDistance(x);
print D;
```

Figure A.6 Distances between Two-Dimensional Points

D		
0	1	2.236068
1	0	2.8284271
2.236068	2.8284271	0

A.9.2   The FROOT Function

The FROOT function numerically finds zeros of a univariate function. The BISECTION module implements some of the functionality of the FROOT function. To use the BISECTION module, the function whose root is desired *must* be named FUNC.

```
/* Bisection: find root on bracketing interval [a,b].
   If x0 is the true root, find c such that
   either |x0-c| < dx or |f(c)| < dy.
   You could pass dx and dy as parameters. */
start Bisection(a, b);
   dx = 1e-6; dy = 1e-4;
   do i = 1 to 100;                             /* max iterations */
      c = (a+b)/2;
      if abs(Func(c)) < dy | (b-a)/2 < dx then
         return(c);
      if Func(a)#Func(c) > 0 then a = c;
      else b = c;
   end;
   return (.);                                  /* no convergence */
finish;

/* test it: Find q such that F(q) = target */
start Func(x) global(target);
   cdf = (x + x##3 + x##5)/3;
   return( cdf-target );
finish;
```

```
target = 0.5;                    /* global variable used by Func module */
q = Bisection(0,1);              /* find root on interval [0,1]         */
print q;
```

Figure A.7 Using Bisection to Solve for a Quantile

q
0.7706299

A.9.3   The SQRVECH Function

The SQRVECH function converts a symmetric matrix that is stored columnwise to a square matrix. The SQRVECH function is used in Section 8.5.2 and Section 10.4.2. The MYSQRVECH function duplicates the functionality of the SQRVECH function.

```
/* function that duplicates the SQRVECH function */
start MySqrVech(x);
   m = nrow(x)*ncol(x);
   n = floor( (sqrt(8*m+1)-1)/2 );
   if m ^= n*(n+1)/2 then do;
      print "Invalid length for input vector"; STOP;
   end;
   U = j(n,n,0);
   col = repeat(1:nrow(U), nrow(U));
   row = T(col);
   idx = loc(row<=col);         /* indices of upper triangular matrix */
   U[idx] = x;                  /* assign values to upper triangular  */
   L = T(U);                    /* copy to lower triangular           */
   idx = loc(row=col);          /* indices of diagonal elements       */
   L[idx] = 0;                  /* zero out diagonal for L            */
   return( U + L );             /* return symmetric matrix            */
finish;

y = 1:15;
z = MySqrVech(y);
print z;
```

Figure A.8 A Symmetric Matrix

z				
1	2	3	4	5
2	6	7	8	9
3	7	10	11	12
4	8	11	13	14
5	9	12	14	15

A.9.4   The SAMPLE Function

The SAMPLE function generates a random sample from a finite set. The SAMPLE function is used for bootstrapping in Section 15.5. The SAMPLEREPLACE module implements random sampling with replacement and equal probability. The SAMPLEREPLACE function returns an $n \times k$ matrix of elements sampled with replacement from a finite set.

```
/* Random sampling with replacement and uniform probability.
   Input: A is an input vector.
   Output: (n x k) matrix of random values from A. */
start SampleReplace(A, n, k);
   r = j(n, k);                           /* allocate result matrix  */
   call randgen(r, "Uniform");            /* fill with random U(0,1) */
   r = ceil(nrow(A)*ncol(A)*r);           /* integers 1,2,...,ncol(A)*/
   return(shape(A[r], n));                /* reshape and return      */
finish;

x = {A B C A A B};
call randseed(1);
s = SampleReplace(x, 3, 4);
print s;
```

Figure A.9 Samples with Replacement

	s		
B	B	A	B
A	B	B	C
A	B	A	A

A.10   SAS/IML Modules for Sample Moments

This section includes modules for computing the sample skewness and excess kurtosis of a univariate sample.

```
proc iml;
/* Formulas for skewness and kurtosis from Kendall and Stuart (1969)
   The Advanced Theory of Statistics, Volume 1, p. 85.
*/
/* Compute sample skewness for columns of x */
start Skewness(x);
   n = (x^=.)[+,];                        /* countn(x, "col")   */
   c = x - x[:,];                         /* x - mean(x)        */
   k2 = (c##2)[+,] / (n-1);               /* variance = k2      */
   k3 = (c##3)[+,] # n / ((n-1)#(n-2));
   skew = k3 / k2##1.5;
   return( skew );
finish;
```

```
/* Compute sample (excess) kurtosis for columns of x */
start Kurtosis(x);
   n = (x^=.)[+,];                               /* countn(x, "col")     */
   c = x - x[:,];                                /* x - mean(x)          */
   c2 = c##2;
   m2 = c2[+,]/n;                                /* 2nd central moments */
   m4 = (c2##2)[+,]/n;                           /* 4th central moments */

   k2 = m2 # n / (n-1);                          /* variance = k2        */
   k4 = n##2 /((n-1)#(n-2)#(n-3)) # ((n+1)#m4 - 3*(n-1)#m2##2);
   kurtosis = k4 / k2##2;                        /* excess kurtosis      */
   return( kurtosis );
finish;


/* for the Gamma(4) distribution, the skewness
   is 2/sqrt(4) = 1 and the kurtosis is 6/4 = 1.5 */
call randseed(1);
x = j(10000,1);
call randgen(x, "Gamma", 4);
skew = skewness(x);
kurt = kurtosis(x);
print skew kurt;
```

Figure A.10 Sample Skewness and Kurtosis

skew	kurt
0.9822668	1.4723117

In many applications, several sample moments are needed for a computation or analysis. In these situations, it is more efficient to compute the first four moments in a single call, as follows:

```
/* Return 4 x p matrix, M, where
      M[1,] contains mean of each column of X
      M[2,] contains variance of each column of X
      M[3,] contains skewness of each column of X
      M[4,] contains kurtosis of each column of X    */
start Moments(X);
   n = (x^=.)[+,];                               /* countn(x, "col")     */
   m1 =x[:,];                                    /* mean(x)              */
   c = x-m1;
   m2 = (c##2)[+,]/n;                            /* 2nd central moments */
   m3 = (c##3)[+,]/n;                            /* 3rd central moments */
   m4 = (c##4)[+,]/n;                            /* 4th central moments */

   M = j(4, ncol(X));
   M[1,] = m1;
   M[2,] = n/(n-1) # m2;                         /* variance             */
   k3 = n##2 /((n-1)#(n-2)) # m3;
   M[3,] = k3 / (M[2,])##1.5;                    /* skewness             */
   k4 = n##2 /((n-1)#(n-2)#(n-3)) # ((n+1)#m4 - 3*(n-1)#m2##2);
   M[4,] = k4 / (M[2,])##2;                      /* excess kurtosis      */
   return( M );
finish;
```

A.11   References

Wicklin, R. (2010), *Statistical Programming with SAS/IML Software*, Cary, NC: SAS Institute Inc.

Index

ACCELERATE YOUR SAS® KNOWLEDGE WITH SAS BOOKS

Learn about our authors and their books, download free chapters, access example code and data, and more at **support.sas.com/authors**.

Browse our full catalog to find additional books that are just right for you at **support.sas.com/bookstore**.

Subscribe to our monthly e-newsletter to get the latest on new books, documentation, and tips—delivered to you—at **support.sas.com/sbr**.

Browse and search free SAS documentation sorted by release and by product at **support.sas.com/documentation**.

Email us: sasbook@sas.com
Call: 800-727-3228

THE POWER TO KNOW®

CPSIA information can be obtained at www.ICGtesting.com
Printed in the USA
LVOW01s0148120814

398655LV00002B/9/P

9 781612 903323